ONE DAY ON BEETLE ROCK

ONE DAY ON BEETLE ROCK

SALLY CARRIGHAR

Foreword by David Rains Wallace

Illustrations by Carl Dennis Buell

SANTA CLARA UNIVERSITY, SANTA CLARA
HEYDAY BOOKS, BERKELEY

A California Legacy book, co-published by Santa Clara University and Heyday Books. For other California Legacy titles or to learn more about the series, see back of book.

Published by arrangement with Alfred A. Knopf, a division of Random House, Inc.

Library of Congress Cataloging-in-Publication Data

Carrighar, Sally.
One day on Beetle Rock / Sally Carrighar ; foreword by David Rains Wallace ; introductory note by Robert C. Miller.
p. cm. — (A California legacy book)
Originally published: New York : A.A. Knopf, c1944.
ISBN 1-890771-53-8 (pbk. : alk. paper)
1. Natural history—Sierra Nevada (Calif. and Nev.) 2. Zoology—Sierra Nevada (Calif. and Nev.) I. Title. II. Series.
QH104.5.S54 C27 2002
578'.09794'4—dc21

2002004423

Illustrations by Carl Dennis Buell

Interior Design: Philip Krayna Design, Berkeley, California

Printing and Binding: Hignell Book Printing, Winnipeg, MB

Orders, inquiries, and correspondence should be addressed to:

Heyday Books
P. O. Box 9145, Berkeley, CA 94709
(510) 549-3564, Fax (510) 549-1889
www.heydaybooks.com

Printed in Canada
10 9 8 7 6 5 4 3 2 1

CONTENTS

FOREWORD

SALLY CARRIGHAR did as much as any writer to propel the wave of interest in natural history and conservation that crested on Earth Day, 1970. I speak from experience. *One Day on Beetle Rock* was a big influence on my first book,[1] to which Carrighar contributed a blurb, doubtless recognizing that imitation is the sincerest form of flattery. But wild nature was much more to her than a literary stock-in-trade. Her 1973 memoir, *Home to the Wilderness*[2] reveals that she integrated her life with her art to an unusual degree, overcoming great obstacles in the process.

Carrighar's upbringing in Cleveland and Kansas City was anything but a tranquil apprenticeship in natural history. According to Carrighar's memoir, her mother, traumatized by a near-fatal delivery, conceived an irrational aversion for her temporarily disfigured daughter, once nearly strangling her. Repeatedly forced to drop out of school because of a stress-related heart condition, Carrighar had an isolated childhood. After two years at Wellesley, the shy, sensitive young woman drifted to Los Angeles and found herself working as a script girl for a silent film industry wherein her employer, Cecil B. DeMille, staged "Roman orgy" prostitute parties for studio financiers. One of her few good Hollywood memories was of a lion, bullied by its trainer, that sensed her sympathy and leapt down from a film set to sit beside her.

Drained by Lotusland's anomie and ninety-six-hour workweeks, Carrighar fled to San Francisco, hoping to become a writer. She had planned to be a concert pianist but lacked the strength, and now her attempts at writing fiction seemed a continuation of that failure. Starving in a rented room, she became suicidal, but a sympathetic psychoanalyst rescued her. She established herself as a radio scriptwriter, then succumbed to a sense of futility and decided to abandon writing. Again, she found herself drifting, alone, in a rented room.

A flock of linnets and a mouse saved Carrighar's sanity. Lying listlessly in bed, she grew interested in the birds' behavior in a tree outside her window and began feeding them. They soon lost their fear of her and even roosted in her room, where she could watch them intimately. One came so close that she could see him watch red clouds drifting over at sunset.

"His eyes would focus on one or more of the pink fluffs, follow it past the tops of the trees, come back and follow others, again and again," she wrote. "He had eyes that saw colors, and were the clouds giving him pleasure?" After the linnets departed to nest, Carrighar acquired an even more aesthetically appreciative companion. She began to hear "birdlike or flutelike" notes accompanying the classical music on her radio—sounds that continued after she turned the radio off. When she moved the radio, a mouse ran out, and when she asked a zoologist if mice sing, he said they occasionally do. The mouse continued to sing in her radio (only to music—dramas didn't inspire it).

When she tried to imagine words to describe the mouse's song, Carrighar wrote:

> I had to get out of bed and walk up and down because of a thought so startling and so exciting: this is what I should write

> about! Birds and animals! I could, after all, be a writer, a nature writer....There could be no finer subject than woods and fields, streams, lakes, and mountainsides, and the creatures who live in that world. It would be a subject of inexhaustible interest, a supreme joy to be learning to tell it straight and truthfully. My whole future life burst open that night like some great and beautiful flower, blossoming in the span of a thought.

One Day on Beetle Rock is the first, and perhaps the freshest, product of Carrighar's "startling and exciting" thought. She spent the next seven years working on the book, educating herself as a naturalist and ethologist in the process and ransacking the Bay Area's science establishment in search of information. When she wanted to know if a predator's scent frightens deer mice, for example, she got Raymond Hall of the Museum of Vertebrate Zoology at the University of California, Berkeley, to try out weasel spoor on some caged mice (it did frighten them). "She comes in here and asks questions I have not thought of before," said Robert T. Orr of the California Academy of Sciences, "and we begin taking down books and [we] discover that no one else has thought of them; so then she goes up in the mountains and tries to find the answers herself."

Carrighar spent four summers in the mountains at Beetle Rock in Sequoia National Park to learn about the creatures whose stories she tells in the book. The park was lightly visited in the late 1930s, and she was able, in a small tourist cabin, to live as intimately with wild creatures as she had with the linnets and mouse in San Francisco. They too became accustomed to her presence and behaved accordingly. She saw many of the things she describes in the book, including an odd association between a grouse and a mule deer. By imaginatively combining her observations of their behavior with her knowledge of the place, Carrighar constructed lively and convincing portraits not only of the animals' life histories but of their psychologies, their mental responses to their habitats and fellow organisms. Her weasel "never quite relaxes, since her longing for sensation seldom can be filled completely." Carrighar's chickaree possesses his treetop territory by "knowing all its turns, its twigs and lichens, and the leaping distance from other bows." Her Steller's jay has an unquenchable urge to dominate his environment by snooping and screeching. Anyone who has watched a weasel hunting, a chickaree climbing, or a jay scolding will recognize her portraits.

Writers had created imaginative portraits of wild animals before, but none had gone so far into their psychology. Ernest Thompson Seton based his lives of big game animals on his extensive experience as a hunter and trapper, but he fell back on a fantasized anthropomorphism to color his narratives. Rachel Carson based her 1941 book *Under the Sea Wind* on her expertise as a marine biologist but concentrated more on imagining migrations, breeding, and other biological phenomena than psychological ones. No author had attempted the minute-by-minute interior account of animals' activities that Carrighar accomplished in *Beetle Rock* as she told and retold the events of a particular day, June 18, from the viewpoints of nine individuals: a weasel, a grouse, a chickaree, a bear, a lizard, a coyote, a deer mouse, a Steller's jay, and a mule deer. Working her careful observations of their foraging, movements, and interactions into the narrative, Carrighar made something new.

Carrighar grew so familiar with some of her characters that they moved in with her after a goshawk grabbed a male grouse in the front yard. "The door of the cabin was open, and in a wave of terror all the rest of the creatures, including the birds, swept in," she recalled in 1973. "I sat among the refugees, a dozen within arm's reach and two on my lap. It had not occurred to me that I would create an emergency by assembling this group at the cabin." Carrighar saved the situation by beginning to feed the creatures inside her cabin instead of outside, as she had been doing. The goshawk eventually got tired of waiting to grab another victim and moved on.

This story, of course, raises a problematic aspect of Carrighar's enterprise. She attracted some of the creatures by feeding them (at risk to herself—a deer kicked her and a squirrel bit her when they felt she was withholding food, and a bear stalked her for weeks, perhaps because she wasn't feeding him). Observations made in a way that so obviously changes animals' lives are scientifically questionable, and as Carrighar learned, there are ethical issues. It looked for a while as though the problem bear would have to be destroyed, although she finally managed to chase him away. Yet, although no park ranger would want a visitor today to feed the wildlife, something of lasting value came from Carrighar's actions. Too often society draws a veil between humans and wild animals, and although the reasons for it are clear, as with Carrighar's unfortunate grouse, the result is a pervasive alienation from the natural world. Carrighar's writing helps to alleviate that alienation.

On Beetle Rock, Carrighar explored not only animals' psychologies, but their implications for humans—an essential exploration if we are to establish a sustainable society. "Many people assume that the true state of nature is anarchy," she wrote in *Home to the Wilderness*:

> That was not what I found, at Beetle Rock or in more remote congregations of wildlife. There were dramas, some very sad,

> and occasionally I was in danger, but what impressed me more were the stability and sanity...here in the wilderness there was a code of behavior so well understood and so well respected that the laws could be depended on not to be broken....Once the wilderness code was ours. Once we were a species that survived in a wild community, among our animal neighbors, because our species too was one of the morally fittest. And when we became more human, when we emerged into the stage of cerebral thought and language, so that we could find words for our moral standards, we did not have to look further for them than our own biological background, our inherited customs and usages. Even before we had religious feelings we must have been moral people. Only later would we have attached those standards, those wilderness values, to our dawning religious consciousness.

With hindsight we can see that the conclusions Carrighar expressed in 1973 may have been a little too clear and bright, arising as much from a strong will as from patient observation. Later observers, Jane Goodall being the best known, have shown that wild animals can be cannibals, philanderers, infanticides, serial killers, and just about anything else that doesn't require human technology to magnify its brutality. But that doesn't necessarily disprove Carrighar's point. "Yes," she might have said, not being one to deny unpleasant facts, having experienced her share, "but those are not usually the behaviors that lead to survival." And I think she would have been right. A lot of what Carrighar records in wild animals does seem moral and sane. Survival may very well depend to a large extent on the code she perceived at Beetle Rock.

Carrighar's insights certainly struck a chord with her contemporary public. *The Saturday Evening Post* carried excerpts from *Beetle Rock* before Alfred A. Knopf published it in 1944, and Walt Disney filmed it afterward. Knopf reprinted it ten times in the next decade. *Harper's*, *The*

Atlantic Monthly, and *Esquire* published Carrighar's writing, and her subsequent books about Wyoming and Alaska were national best-sellers. Through the fifties and sixties, Carrighar was as widely read as Rachel Carson and Aldo Leopold, and although she did not receive the same attention from academia and the media, her highly readable books were a ubiquitous part of the 1960s environmentalist wave.

Later, this very popularity began to undermine her influence. Universities and environmental groups maintained Carson's and Leopold's reputations while trade-glutted bookstores and libraries let Carrighar's slip away. And her subject worked against her academic reputation. Carson and Leopold wrote about wildlife ecology, a concrete discipline which lends itself to institutional agendas. Carrighar wrote about wildlife behavior, and beyond that, wildlife consciousness, which institutions have trouble addressing. Some academic philosophers and psychologists still doubt that nonhuman animals have consciousness (as though they've never known a dog or cat), and the imaginative exploration of the nonhuman mind that Carrighar attempted remains unconventional.

Carrighar's approach to nature became unfashionable, as did Rachel Carson's. Both women downplayed their personalities while concentrating on nonhuman subjects. Both were agnostics, exploring nature without reference to traditional ideas of spirituality. (Carrighar thought she might have experienced telepathic communication with animals, but she didn't seek spiritual explanations.) In the era after World War II, this down-to-earth, impersonal approach appealed to the expansive sense of discovery that pervaded mainstream society. After 1970, as the Vietnam War and other factors fragmented the mainstream, nature writing became more subjective. Most popular writers played up their personalities, and many approached nature through traditions of spirituality. Carrighar and Carson seemed bland to readers attuned to personal revelations and transcendent intuitions, although these sometimes esoteric features yielded works with a far narrower appeal than the best-selling nature writing of Carrighar's time.

Books with an impersonal, "nondenominational" approach to nature were displaced by television shows. Well-funded documentary teams could offer audiences much more spectacular glimpses of wildlife behavior than a lone writer like Carrighar could present. Yet although the likes of David Attenborough do fine work, documentarians have trouble presenting their material with the strength of an imaginative sensibility like Carrighar's. Scriptwriters and editors often reduce wonderful footage to hackneyed vignettes: "See graceful antelope; see fierce lions; see lions stalk antelope; see antelope flee lions; etc." Carrighar's animal stories follow some of the same patterns, since they are natural ones, but Carrighar's are too closely observed and deeply felt to be hackneyed. They also are better told. Few television scriptwriters have the literary skill to tell nine separate versions of the same day's physical and mental events. Few have the talent to relate atmosphere and place to animal behavior as she does:

> In the trees' fluff of shadows that night were many small, bright, black points, the eyes of birds wakened by the conflict of winds. The winds reached the coast as a single north-flowing stream, but the Sierra ridges divided it. Above Beetle Rock two of the currents were meeting again. One had swept down from Mineral King, unopposed until midnight; then the canyon wind strengthened, rising against the slope. The colliding gusts battered the trees.

> Birds like the flickers and chickadees, in the hollows of rigid dead snags, were not roused, but the wind alarmed every bird on an outside perch. It was giving a wild ride to the Steller's jay clutching the end of a limber branch. And no bird could fly to a firmer roost until morning, since wings cannot grope their way through the dark as feet can.

Few have observed animals as sympathetically and intelligently, and few have described them as imaginatively and eloquently as Carrighar did in *One Day on Beetle Rock*. Were the experiences she attributes to her wild creatures real? It is largely impossible to say, of course. We know more about how animals behave than we did in her time, but we know little more about how they feel and think. If we are at least more willing to acknowledge our ignorance now, her writing has had a lot to do with it.

— DAVID RAINS WALLACE
March 2002

1. *The Dark Range: A Naturalist's Night Notebook*
by David Rains Wallace (San Francisco: Sierra Club Books, 1978)

2. *Home to the Wilderness*
by Sally Carrighar (Boston: Houghton Mifflin, 1973)

INTRODUCTORY NOTE

THIS IS A DANGEROUS BOOK, full of disturbing possibilities. Should it fall into the hands of the young, it is extremely likely to make naturalists of them. Even a hardened adult must read it at his own risk—the risk of being seized with an overwhelming desire to hear the wind in the treetops and to smell the incense of the forest; to watch a lizard sunning itself on a rock, to glimpse the lithe form of a weasel disappearing over a log, or to come briefly face to face on the trail with a startled buck; to hear the evening songs of birds, to see the bats come out at dusk, and to share with the creatures of the wilderness the adventures of the night.

He may even have to visit the particular place in Sequoia National Park which is here so accurately, compellingly described, and see with his own eyes its animal inhabitants leading their busy, interesting, self-sufficient lives.

These are stories of the adventures of animals, but with a difference—the stories are of actual animals in an actual place, as the author has observed them. She has watched carefully and reported truthfully, always with sensitive understanding and a keen awareness of beauty. The tales are fiction, yes, but fiction closely parallel with fact. This is real natural history.

—ROBERT C. MILLER
Director, California Academy of Sciences
(Introductory Note to the 1944 edition)

BEETLE ROCK

THE WATER OF THE BROOK reflected the sunlight up to an alder branch, where it flickered along the gray bark. On the fool's gold under the ripples lay a web of the sunlight, gently shaken. The sound of the current was subdued here, and the stir of the streamside leaves, and the Mule Deer was quiet too, as he moved slowly forward along the bank, clipping off willow buds. But the Deer was approaching a rockier channel in which the stream tumbled and splashed. When he reached the cascading water, suddenly he leapt over it, and back again, and then stood in the blowing spray, sharply white and cold, with his head flung up and excitement in his eyes.

In April all the bucks had a wild playfulness, quick to rise. Two would be sniffing for acorns among the dry leaves beneath an oak, when one would spring away—a challenge, and the second must follow. The deer's hoofs would make their clean arcs over the yuccas, over the redbuds, over the mounds of yellow fremontia blooms, arcs so high that other deer could have stood under them. At the end of the chase, the bucks would meet in a climax of vertical leaps, tossing heads, and whistling breaths. Afterwards some watching fawn might try bounding over a small yucca or redbud, but the mood of the older bucks seemed the more resilient.

Were they feeling only the exuberance of spring, or did the older deer have a memory of Beetle Rock, and were they pleased to be climbing again to the granite field in the sky?

The herd had wintered low on the ridge between the Middle and Marble Forks of the Kaweah River. They had been down in the dwarf forest, where the same storms that heaped snow on Beetle Rock fell as rain, and had brought up fescue and wild oat grass. Now the mountain's snow cover was shrinking upward. The deer stayed below the edge of it, since snow would hinder their flight from coyotes and cougars. They grazed back and forth on the side of the ridge, rising about a thousand feet every month.

The ridge became steeper the higher they climbed. In many places they had to circle sheer granite slides, polished by ancient glaciers. Beetle Rock was a similar part of the mountain's skeleton, protruding through soil and forest. But the Rock lay at the crest and its top was nearly as flat as a meadow. By summer the upper surface would be so airy and warm, with so many crannies for hiding, and such an abundance of food, that more than forty animal species, besides the insects, would be competing for homesites there. Down in the chaparral the deer had left a group of winter companions. Upon the slope they were meeting others. But nowhere in their wanderings did they find such a dense population of animals as in summer at the Rock.

By April the herd were a mile above sea level. Even there the trees were not much more than brush-high. Looking out over them the deer could see the series of mountain ranges that piled up to the Western Divide. Dropping away on the other side was the Marble Fork canyon, a vast tapering cut, spreading out at the low end upon the floor of California's great Central Valley, narrowing at the upper end between towering ridges. Beetle Rock, directly above the deer, was still covered with snow, still only a short, smooth line breaking into the treetops along the sky.

Abruptly the deer found themselves beneath firs and ponderosa pines, a few weeks later beneath sugar pines and the giant sequoias, some of the tallest trees in the world. Trees like these formed a wall around Beetle Rock's inner edge. Back among them were cabins for human beings, looking very small; six or eight of them could have been built upon one another without reaching to the lowest boughs of many of the trees. Some of the pines were twice as tall as the same trees elsewhere. Here they had a longer growing season, and perhaps the trees responded to the light on these Sierra Nevada mountains. The subtropical light, so clear on the valley below, was even more fresh and elastic at the Rock. The atmosphere seemed light more than air above the milky stone. Had the trees thrust their tips up as far as possible to reach more of it?

The deer came to Beetle Rock on a morning in May. The herd Buck, the leader, took them around the cliff's semicircular base to the west side, then back into a green-shaded gorge. Along its bottom a stream, now at flood stage, was thundering towards the canyon; before the deer started down the mountain again it would become a trickle. The herd's upward migration was over, for the meadow where all of them had been born (all but one buck) lay only a few bounds off the top of the left slope. One by one the deer leapt the water to climb to it. But the leader turned right. He would go up on Beetle Rock first. Among the snow patches on the side of the gorge he found the animal path he would follow daily, now, for the tenth summer.

The Buck passed a dogwood thicket where he was accustomed to lie and chew his cuds. He continued up through an open grove with scattered manzanita and chinquapin brush among the trees. Hearing the cry of a bear cub, he bounded behind a clump of the manzanita. From that shelter he peered out, and farther along the slope saw a creature he knew well, the Black Bear. She would be irritable this year, for she was guarding two young ones. On this morning they were having a lesson in tree climbing. A black cub had tried to go up a cedar, but his claws were pulling out of the fibrous bark. It was not the slide down the trunk that distressed him, but the coming rebuke for forgetting that bears do not climb cedars. On the ground he galloped away from his mother, but her paw caught him and sent him rolling. A brown cub, pigeon-toed in both front and hind feet, was shambling towards a pine trunk.

When the Buck came to the Rock's corner, he found a brook flowing down along its north side. As soon as the snow was gone, the brook would be dry and its bed would become the gulch, the draw, used by many animals as the shortest route from the slope to the top of the Rock. The Buck walked on up beside the water.

He saw the Weasel slip out from a chink between stones. A limp deer mouse hung from her mouth. With bold, arching leaps she crossed a log over the brook; had she no enemies that she, so small, could be so arrogant?

She scarcely had reached the log when the mate of the deer mouse appeared in the hole. She paused for a wary instant, then darted out, carrying one of her tiny young. She would try to take all her brood to a new hiding place before the Weasel returned, but one, at least, would be safe. Through the summer the Buck would see that small Deer Mouse grow and establish her own family.

The Buck reached the fir tree, on the forest side of the draw, beneath which he often rested on summer afternoons. Now down from its branches somersaulted the Chickaree, vanishing, saucy chatter and fur, in a snowbank. Quickly he sputtered out, snatched a twig from the ground, leapt on the trunk, and was bounding away through the tree, probably towards a half-finished nest. In the Buck's eyes was a flicker of recognition. The squirrel, almost grown now, had been the liveliest of a family reared the previous year in a pine stub near the draw.

The Buck looked back into the thicket behind the fir, back through the bare trunks of the trees, seeming to crowd closer as they receded, the bark of the firs gray, of the pines wine-color, of the sequoias rust. The trunks seemed to lift their masses of overhead needles up even among the white cloud puffs motionless in the sky. The Deer's eyes did not search the trees, but rather exposed themselves to movements there. His gaze was passive until he discovered the Steller's Jay, hidden within a dense fir, turning its head to search in the bark for grubs. Few eyes but a deer's would find the Jay during this month, when his sense of responsibility for his young made him one of the most secretive Beetle Rock birds.

Even the Buck had not seen the Grouse, although she was less than a deer's length from him, concealed beneath a seedling pine. Now she came out to lie in the sunlight, and startled him, for she did not move with a grouse's usual, unobtrusive step. She fluttered, half flying, half hopping. The Buck sensed that she was wounded or sick. Her eyes were sunken, and her feathers were dull and disordered, showing weakness as plainly as human pallor does.

The Grouse stopped at his very feet. She may have been one of the last year's young, for he did not remember her. He stayed with her under the fir for a while. He too lay upon the needles, chewing his cuds, watching the tired lids close over the Grouse's eyes as if she felt safer with him there, depending on his alarm to warn her if an enemy should approach. When he rose to go over on the Rock, she fluttered back into hiding.

Several rills joined to form the brook, a short distance above the fir. The Buck waded across them, and finally stood on the inner edge of the Rock. Before him lay the two acres of granite field, almost square, divided like a beetle's back. The forest rose behind the north and east sides. The other sides ended in a sharp rim out over the canyon.

Most of the Rock's terraces were smooth, but some were broken up into domes, knobs, and angular boulders. A few pine and cedar trees had found rootholds in the crevices. Their shade lay cool on the gray and ivory-white stone. Brush was growing from soil blown into the cracks; rock flowers and grasses leaned from smaller pockets; back under the ledges were cliff-brake and moss. On the May morning when the Buck came to the Rock, a green fuzz was showing in some of the shallow depressions. A month later golden-throated gilia and baby's breath would be blooming there.

Energetically the birds were singing their claims to their particular trees and bushes, and small furred creatures were chasing each other, settling the boundary lines between their private areas of the Rock. Before the Buck had gone many steps he came to a bluffing contest. The Lizard and a neighbor who coveted his gully were pumping up and down on their forelegs, trying to intimidate each other by displaying their bulging blue throat-spots.

The Buck moved from one terrace to another, browsing on the manzanita bushes, straying towards the north side of the Rock. The deer did not often go to the south side, for there they found a disturbing bear scent. It rose from a small meadow below, and from a wooded shelf on the face of the cliff where several bears stayed in the daytime. This morning the Buck caught none of their odor. The air was lightly pungent with the fragrance of new needles on the trees and new leaves on the brush. And everywhere was the sound, which never ceased here, of the canyon wind flowing among the crevices, turning between the boulders, blowing upon thin edges of stone, the wind that had given sound to the mountain before any leaves grew there to rustle, the wind that the Buck would not hear forever.

But now he has leapt to a higher ledge, is ready to bound away from the Rock, and into the forest. For a turn past a chinquapin has disclosed the Coyote, stalking granite-hoppers in a space hidden by brush. The Coyote sensed that the Buck could escape, and gave no sign that he heard the scattering gravel and the hoofs' landing upon the ledge. The Buck looked down from above. The Coyote was new here; for three years no coyotes had lived in the deer's immediate range.

This summer, then, the herd leader would need even more than his usual caution. Wariness was his skill, his most significant quality. In the Buck was the best example, among all the Beetle Rock animals, of the willing tension that keeps a wilderness society stable. His was the finest alertness, but every creature had much of it. Since long before he was born, the community here had been holding together, because each of its members was ready to leap, to chase, to freeze, to threaten, to love, or to step aside—in an instant.

WHAT HAPPENED TO THE WEASEL

NIGHT'S END HAD COME, with its interlude of peace, on the animal trails. The scents that lay like vines across the forest floor were faded now, and uninteresting. Hungry eyes had ceased their watch of the moonlight splashes and the plumy, shimmering treetops. No heart caught with fear when a twig fell or a pebble rolled. For most of the nocturnal hunters had returned to their dens, or ignored one another in a truce of weariness.

From the frail defense of an oak leaf a deer mouse stared at a passing coyote, sensing its safety by the mechanical tread of the great paws. A frog and an owl at opposite ends of the same tree closed their eyes. A black bear, trampling a new bed at the base of a cedar, broke into the burrow of a ground squirrel. With heavy eyes he saw it leap to a rock-pile; then he made a last slow turn and curled himself against the trunk.

The Weasel was not tired, and never joined a truce. She was stung by only a sharper fury when she saw the darkness seeping away beneath the trees. On the hillside where she hunted with her young she suddenly pulled herself up, sweeping the slope with her nose and eyes, trying to cup the forest in her ears for the sound of a chirp, a breath, or an earth-plug being pushed into a burrow. There was silence—proof that all the quick feet had been folded into furry flanks. She and her kits were alone in a deserted world.

The Weasel too was leading her family home, but she had stopped to try to stir up one more chase. She had chosen a slope that never furnished much excitement. The ground was a clear, smooth bed of pine and sequoia needles, with no underbrush where victims might be hiding. Even the odors beneath the Weasel's nose were of little help. For here no large obstructions, no fallen logs or gullies, had gathered the scent threads into strands. Still she whipped across the surface, vainly searching. It was not that she needed food after the night's good hunting. She was a squirrel's length stripped to a mouse's width, and was no glutton. But she was driven by insatiable hungers of the nerves.

Now she has caught the scent of a chipmunk, redolent and sweet. Perhaps it will lead her to the chipmunk's nest. She bounds along the path of odor with her tense tail high. But here is the trail of a second chipmunk crossing the first. The Weasel stops, confused. Now she follows one trail, now the other. Back and forth across the slope, the odors weave a record of two chipmunks chasing each other. But where are the small warm bodies that left the tracings of delicious fragrance? The Weasel turns in her own tracks, comes to an angry stop. Her five young watch her. What will she do now? She'll forget the chipmunks. She stands erect, moving her nose through the air as she tries for a different scent.

Her nostrils trembled with her eagerness to find an animal odor in the smell of needles, loam, and cool dank funguses. She caught the juiciness of crushed grass mixed with faint musk. Meadow mouse! Off again, she sped along the mouse's trail towards the stream below. But the trail suddenly ended in a splash of mouse's blood and coyote scent.

The intense hope of the Weasel snapped into rage. The young ones saw her swirling over the needles like a lash. If there was another scent trail here she'd find it. She did—at this blended musk and pitchy odor left by a chickaree when he jumped from the trunk of a pine. The odor line turned to a patch of cleared earth, where he had patted down a seed, and then to the base of another pine, and up. The Weasel pursued the scent to one of the higher branches and out to the tip. From there the squirrel had leapt to another tree. That was an airy trail no enemy could follow.

The Weasel came down the tree in spirals, head first, slowly. When she reached the ground she paused, one forefoot on a root. Her eyes looked out unblinking and preoccupied. Perhaps her hungers were discouraged now—but no. Her crouched back straightened, sending her over the root in a level dash.

The weasel young had scattered while their mother trailed the squirrel. They came flying back when a high bark told them that she had made a find at last. She was rolling over and over with the body of a chipmunk. This was not like her usual, quick death blow; again she drove her fangs through the chipmunk's fur. Then the harsh play ended. The Weasel leapt aside, allowing her kits to close in on the quiet prey.

While the brood fought over the chipmunk, their mother ran across the slope to explore the leaves beneath a dogwood thicket. By the time she returned, the shadows were thin and the chill of dawn was creeping in among the trees. Two of the young weasels munched last bites, but the others moved about slowly, only half alert, their tired legs hardly lifting their bodies above the ground. The mother bounded in among them. Her own strength still was keen but the kits needed rest, so she called them and the little pack moved down the hill.

At the base of the slope they must cross the stream. An uprooted sugar pine leaned from one side and a silvered fir snag from the other, making a bridge with a short gap in the middle. A few times when the kits were smaller, one had missed his footing and had fallen into the water, but this time, tired though they were, all made the jump with safety.

The weasels' den was in a thicket, a few bounds off the top of Beetle Rock. To reach it they climbed the slope beyond the stream. When the Weasel approached the cliff from below, she often circled north and up through the brush at the end. Now she led the kits home the short way, over the Rock's broad, open terraces. They met no other animals until they came upon two gray mounds, strong with human scent. The Weasel dodged into a crack between the granite slabs. By connecting crevices she evaded the sleeping human forms and brought the kits to familiar ground beneath a shrubby oak. There, one by one, the six small creatures slipped into the earth.

A CRY FROM THE STELLER'S JAY reached the Weasel. It was like a touch on a spring. Out of the den instantly, she stretched herself up, her eyes glittering, darting everywhere, trying to see what enemy of the bird was making him so angry. There was a rush of wings as four more jays flew in to aid the screamer. The Weasel often stirred commotion similar to this. She was not the one who had done it now, but if there was going to be a fight she wanted to see it and if possible to join it.

The jays grew even more excited, for they had lost the object of their hunt. They tossed themselves through the branches of a pine, peering everywhere over the Rock. The Weasel found what they sought—a wildcat crouching under an edge of the granite. A figure of wonderful tense stillness was the cat. Watching her, the Weasel swayed with emotion, her ears, her eyes, her pointed tail, her very fur electric. She saw the wildcat wait, quiet but alert, while the voices of the jays grew less and less belligerent. Finally they ceased. When there had been no outcry for some time, the smooth round body slipped from its crevice, into a gully and out of sight. A moment later it appeared at the gully's end, a shadow sliding across the gravel to the brush.

The Weasel flung impatient eyes around her. She was charged with energy and once more had no chance to spend it. The human creatures had partly risen during the jays' noise. Now they were lying down again, composing themselves, but the Weasel could not escape so easily from her emotion. She prowled over the ground, sniffed for a promising scent, but found none. Still restless, she withdrew into the burrow, to lie with her body in the tunnel, her little triangular head at the entrance where she could watch for something to battle or to chase.

The lustre of the moon was dulled by the gray of the dawn, and no pocket of full darkness now remained, not even in the underbrush or the depths of the trees. A few chirps rose from high nests, but as yet no pewee droned, no chickaree's query rang from the pines. After the jays ceased screaming, most of the sounds were lifeless—a small wind knifing through the oak leaves, and the pricking of stiff needles against bark.

All around the Weasel, animals were sleeping. The hunters of the night were drawing long breaths from tired flanks, their eyelids nerveless and their faces empty of expression. Other animals were breathing lightly now. Soon they will waken, stretch, rise, suddenly leap for a fly or perhaps pounce on a leaf mistaken for a mouse. They will feel it good to have their muscles speeded by an inward fire, just as the Weasel does. For them, the fire will burn out later; then they will be glad to lay their tired legs on their beds. Each satisfaction in its time, and no resistance to the loss of any. For the Weasel, though, the fire will not die. Her nerves cry always for more and more intensity—for wilder winds, for colder air, for faster streams, for sharper scents! In the wilderness most animals' needs are answered, and strain is balanced by repose. But the Weasel never quite relaxes, since her longing for sensation seldom can be filled completely.

Her taut nerves had been stretched still tighter by the restricted life she had led while her kits were small. Through the rest of the year she gained some peace by being always on her way to a new place. There were boundaries to the territory where she wandered, but she never returned to a meadow or grove or granite field until the season had moved along and given it a different look and to some extent a different population.

Quick roamer that she was, she found her food by covering much ground lightly. She cleared out the slow, the weak, and the injured prey, and then went on. By June eighteenth, however, she had stayed at Beetle Rock for thirty-seven days.

Long since, she had found the easier victims, and now was forced to hunt with a sickening persistence. The very sight of the Rock—hard, gray, unmoving—was like pressure on a bruise.

The Weasel's den, too, was now almost uninhabitable. It had been a good one this year. The Weasel had got it by eating the gopher that had dug it. The central chamber already had been prepared for a brood of young, with a lining of grass and roots, to which was added gradually the fur of mice. Six side burrows branched from the nest compartment. The tidy Weasel used them for discarding refuse. The entrance had been well built, hidden at first beneath a crust of rains-tiffened leaves. But the crust was broken down. The whole burrow was going to pieces, and the nest was becoming intolerably crowded as the five kits neared full growth.

The dawn on June eighteenth was colorless and cold. Above the canyon staggered a bat, its flight abrupt and senseless to an eye that could not see the insects it pursued. A late owl flapped to rest. The breeze drew across the Weasel's nose a tassel of scents, but none of them interesting: the bitter odor from an anthill, the dried blood of a deer mouse slain by the owl, and smell signs left on a bear-tree. The Weasel noted these and remained where she was. Then she caught another scent, from the black oak overhead. It was faint but exciting—the clean, sweet spiciness of purple finches!

A flock of the birds had flown to the tree on the previous evening, while the Weasel was away. Now they were stirring for a farther flight, beginning to layer their feathers and to stretch their wings. Cautiously the Weasel crept up the trunk of the oak and stepped onto one of the branches. The canny finches were perched at the outermost ends. Twig by twig the Weasel approached them. The branch is bending a little, now, under her weight. Two of the sleepy birds are almost in her grasp. But she cannot spring; the bough here is so light that she must coil her body around it to keep her balance. It dips. The birds are warned and the whole flock bursts from the tree.

The sharp fights of the earlier night, and the prey that had filled the stomachs of the Weasel and all her brood, were too far past to be remembered. She only sensed that she was tortured with disappointments. She stood at the entrance of her burrow and looked again around the Rock, empty of life except for the sleeping human figures. Finally, more in disgust than fatigue, she turned into the tunnel and the nest.

While the weasels slept, the sun rose and became warm, and the daytime animals began their frank and visible play. Only three or four bounds from the den, chipmunks raced on the Rock, the Lizard searched for insects, a junco sang, and a butterfly dangled above a clump of golden-throated gilia. A chickaree danced through the branches of the nearest pine, continually calling in his high, sweet bark.

There was a little time of wary silence when the human creatures threw off their blankets, stood up and stretched, and walked to the rim of the Rock. But they paid more attention to the canyon view than to the scores of eyes that watched them from the pines, the manzanita, and the nooks in the granite. As soon as they took up their blankets and strolled towards the bordering forest, they were followed by a wave of small bright sounds.

The weasels, down in the cool dark earth, lay with their bodies coiled together, breathing softly into one another's fur. But their graceful sleep was jolted. The nest was shaken by a series of great thumps, slight at first but quickly becoming violent.

Two of the smaller weasels were so exhausted that they did not waken, but the mother and three larger kits streamed out. They saw the white rump of a deer who was bounding down the slope, alighting each time with a force that would have collapsed the den if he had struck it. His passage was a crashing—the earth resounding, fallen branches snapping under his feet, and gravel rolling.

The animals' calls and movements all had ceased. But after the deer had disappeared and the forest had been empty of sounds for a moment, a chirp was heard from an oak tree and a chickaree dropped a cone from the top of a pine. Among the leaves a wing flashed and a fluffy tail unfurled. Soon the creatures were frisking more recklessly than ever, relieved to feel again that they were safe.

A golden-mantled squirrel made the mistake of jumping over a log without stopping on top to see what was on the other side. He came down at the mouth of the weasels' den, in the very midst of the four who had awakened. The squirrel had seen them before he touched the ground, and had twisted his body so that he was able instantly to dodge back towards the Rock. But the mother Weasel cleared the log on his heels and caught him.

The two fought back and forth across the granite. The squirrel was larger, the Weasel faster; in fury they were matched. Every breath of the squirrel was a shriek of rage and protest.

In this battle the Weasel must really spend herself. Now the two are together, now apart, streaking away side by side. Together again, they claw and hiss feverishly; their teeth flash and grope for each other. The Weasel has caught the squirrel around the haunches. All her attention is given to coiling herself around his body. She shifts her hold until it is more and more secure and her teeth can approach the base of the squirrel's skull. Finally she sinks them in a precise and fatal puncture.

The feast is shared by all the family except the two still sleeping in the den.

Now the Rock is indeed quiet. Gone are the junco's song, the chickarees' barks, and the quick brown scallops of the chipmunks bouncing among the boulders. No jay screams. This is one of the times when the undependable jays have failed to call a warning. If they saw the flight of the golden-mantled squirrel, and its end in the whirl of cinnamon and yellow fur, they watched without a flutter or a sound.

THE FIGHT with the squirrel unbound the Weasel's nerves and filled her with a sense of ease and grace. For a short time now she found it a pleasure merely to be alive on this height above the canyon, circled with sky and sparkling with pine-sharp light.

Without returning to the sleeping kits, the Weasel led the others down over Beetle Rock to the stream below. She crossed the granite flats with airy leisure, looping high, like an elastic measuring-worm that sprang from the ground each time its two ends came together.

At the base of the Rock the slope was covered with boulders, chaparral, and seedling pines. The Weasel's lithe little body flowed among them. Smoothly she wove her way through the stubs protruding from a fallen cedar, jumped over and under the manzanita branches, and ran along the boughs of the pines. Gravel dust coated leaves and needles, but she melted through without getting a speck of it on her coat. The three kits followed closely. They were not so

agile, and they could not finish a bound with their mother's perfect balance, but they tried and they were learning. Finally the family reached the taller tangle of green along the stream.

On the bank above the water, under a slanting black oak, was a clear space free of brush. Here the three young weasels stopped and looked around, sniffed across the moist earth, jumped on the trunk of the oak and down as if they knew this place and were glad to come back to it. They started to chase each other. The leader leapt upon the tree and then, as the other passed, he bounded to the ground and into a pile of dead leaves. The pursuer whirled and also darted into the leaves. The leader returned and paused, moving his nose deliberately along a root. But when his brother was out of the leaves and streaking towards him, he spun around and caught him. The two became a single boiling mass of fur, paws, tails, and teeth.

While one was on his back, the other astride his belly, the third kit, a female, ran up and took a nip at the victor's leg. Her brother sprang off and after her. He gained as they dashed across the clearing, but just as he would have leapt at her, the other brother streaked between them. The two males locked together, rolled over and over, then suddenly were apart, racing after each other. Their golden bodies were like the darting sunbeams under shaken leaves.

Their game was sharper than the play of most young animals. It was more subtly timed, more finely fought, exquisitely close to deadlier emotions.

The mother had climbed up into the oak tree and watched the kits from the crotch of a branch. Her eyes flicked also to the jerking head of a kingfisher on a higher branch, to the zigzag of a dragonfly, to the water's lacing and the dipping of a mountain lotus. She heard each touch of bough on bough, each gulp of the current. Even the roundness of the branch, the movement of the stream-stirred air, and the smell of the moist bark came to her as brilliant and provocative sensations. Her head swayed on her long neck, now from side to side, now up and down, as she tried to see and smell more, even, than a weasel's senses have been tuned to catch.

Into the amiable, easy morning came a little heaviness. The Weasel dropped from the tree with a silky leap, and down the stream bank to a wet root. She sucked a long drink from the water, then thrust her head entirely under, where the lovely movement was so like her own. Abruptly she pulled up her nose and watched the splash of shining drops. Again she shattered the water, as if she had learned a new game. The young ones, tired of their own game, tumbled down the bank and joined her. Even the leopard lilies at the edge were not more vivid than the tawny splashers.

The Weasel lifted her wet nose high. She found that the air had become cold. The sun was gone, and the leaves above the stream were troubled with a dark motion. Smelling a storm, the mother sprang up the bank and growled to summon the kits. But they were lost in their play. She called again and one of them bounded to her side. The remaining two continued to chase each other in and out of the muddy roots. But when the stream was splashed with heavy raindrops, even the kits were prompted to look for cover.

The weasels raced up the slope towards their den, but the storm was faster, and the family was caught in pelting rain. The mother led the kits up a dead white fir to a hole a dozen weasel-lengths from the ground. They entered and took shelter in the hollow trunk.

Thunder crashed. It brought the mother out to the hole. There she crouched, beside herself with emotion at the violence around her. With a roar the rain streamed down, now flung aslant as the wind tore through the trees, now tumbling upon the forest with seemingly the weight of a cataract. The branches churned and pitched. Some snapped and fell, catching on other boughs, then breaking away and plunging to the ground.

The water had its own smell. More pungent were the moistened odors of needles and bark, of the earth and its decaying cover, of wet granite, and of all the animals, and their trails left everywhere that feet had passed. Soon the air itself was stinging with ozone generated by the lightning. The Weasel sharpened her nose to suck each thrilling scent.

Green flashes fill the forest. They seem to come from the very depths of the trees. In the leafy caverns never reached by sun or moon, the glare exposes the hidden nests of birds and small furred creatures. It points to the mottled owls attempting to hide themselves against the mottled bark.

As the canyon trembles with the shocks hurled on its sides, the Weasel in the fir hole weaves and sways. This at last is a weasel's forest—lights, sounds, scents, intensified to the limit of excitement.

Too soon the storm began to pass. The Weasel felt its going. She detected the first crumbling of the violence—the delay in the thunder, which then broke slowly, lacking its earlier, full, murderous crack. It rolled out longer, a sustained, reverberating roar.

The lightning flashes came less often, and were paler, weaker, a suffused white-green. This was only a pretense of danger. The rain was ceasing; in diminishing waves it was departing down the mountain. Finally there was only water dripping from the leaves.

A break in the clouds showed high blue, and below, in the canyon, sunshine tinted a misty feathered slope. The storm was over. The Weasel's world had fallen away.

She sank within the hole. But soon she reappeared, to lead the three kits down the tree, through the steamy thicket, and over the wet loam to the den. She had not been concerned about the two she had left asleep, for other storms had proved that rain did not drain in. But she uttered a shrill cry at her first sight of the burrow. Over it crouched a great fur body, the Coyote digging to the nest.

His paws worked fast, scattering clumps of mud to both sides. His haunches twitched, and his nose moved impatiently across the fresh-cleared earth. The tiny Weasel instantly attacked. She sprang for the face of the Coyote, where she caught with her teeth through his lower lip. The Coyote yelped. He swung his head to try to shake the Weasel off. Still she clung. Desperate, he gave his head a great jerk and the Weasel was hurled to the ground with his torn flesh in her mouth.

She leapt again for his face. But he was ready; she barely missed death between his snapping jaws. Again she tried. This time she felt the grazing of his teeth. Hopeless now, but wild with grief, she crouched in the hollow center of a log, out of his reach. The Coyote resumed his digging, his sidewise glance on the little head, all fangs and scarlet tongue, and glittering eyes.

The two young weasels were still in the nest when the Coyote, with a last quick scoop, uncovered them. Each was soon quieted, as their mother's cries tore through the soft bright air. The Coyote was deliberate at his meal. When he finished, he walked off slowly across the Rock. The Weasel watched him disappear below its rim.

She continued to stare at the cleft where he had vanished until the frenzy died from her eyes. Then she came down off the log, her body but a loop of limp flesh. She growled to call her three remaining kits, who had hidden in the thicket, and led them to a pile of boulders at the end of a gully. In its crevices she saw them curl themselves for sleep. And then the Weasel too lay down, with her tail across her face.

THE FAMILY did not waken until after dark. Very warily the mother slid from a chink in the rock-pile. She moved up over the stones, from one to another, stopping to trace the air for scents.

The moon had not yet risen, but the stars spread a glow on the granite and turned the canyon mists to pearly white. Above the rock-pile, needled branches of a young fir tree stretched black against the fiery blue of the sky. They were a canopy of darkness leading back to the deeper shadows of the forest.

Now that the weasels' den had been destroyed, there was no reason for staying longer at Beetle Rock. In any case, the kits were nearly old enough to start their roaming. When the sun next rose, the Weasel would not be there to see it shine on the changeless granite. For her, tomorrow's sunlight would be only a brightening of the green shade in the world behind the fir. She barked impatiently to call the kits.

The Weasel had a general memory of her larger territory. She knew where the rivers were, the cliffs, the groves of great trees, and the meadows thick with grass that hid a swarm of mice. Towards one of the upper meadows, now, she led her young.

On every side, in the trees and the earth, the Weasel was aware of flying, creeping, scurrying, digging creatures. She ignored them, and climbed the mountain with a long and steady gait. Most of the trees were sequoias and sugar pines. Above their giant shafts the sky was lost, and the darkness on the ground was so complete that even the Weasel's eyes were of little use. She was guided by the trickle of a brook that flowed, as she remembered, down from her marshy meadow.

Finally the weasels crossed a low ridge. Beyond, the earth dropped to a level where the great trees opened to encircle a huge nest-shaped clearing filled with moonlight. In the quiet air no blade of grass was stirring, no bluebell on its thread of stem, no pine or sequoia branch. The moon, climbing away from the trees around the meadow, seemed the only moving thing. But animal odors were as heavy as a haze.

They told the Weasel that many creatures were secretly active, that a mole was digging a new side-tunnel somewhere near, that a gopher was crouching in its burrow-entrance, waiting for courage to move out half its length for a lily stem. A thick scent rose from the mice—mice eating away the grass to form new runways, divining where seeds were sprouting and scratching away the soil, mice scurrying beneath the arching stems, or sitting motionless on their haunches feeling the good seeds start to digest, while their eyes, grave with little dreams, looked down the moonlit avenues of their grassy world.

Already the Weasel had forgotten her barren days at Beetle Rock. Tonight she had come to a new phase of her year's existence, and she was entering it with all her accumulated strength. As the summer would pass and the kits grow larger, faster, and more intense, her energies would find more and more release. By the time the weasels' fur turned white in the fall, the family would be ready to separate. The first snow would be due, and the mother's life of solitary freedom, as keen and brilliant as the winter sunlight, would commence.

The trunk of a fallen Jeffrey pine stretched towards the center of the meadow. The Weasel sprang upon it and led the kits to the end. Around them the grasses, nearly as high as the top of the log, were a glistening sea into which the Weasel would dive as soon as a trembling showed that a mouse was moving at the roots.

The three kits jostled each other for positions and began a little play. The mother ignored them. On the end of the log she waited, with her eyes fixed on the grass. Poised for a leap, erect, she was as sharp as a small, arrested flame.

WHAT HAPPENED TO THE SIERRA GROUSE

WHILE THE SIERRA GROUSE had been going to sleep on the previous evening, she had watched a large brown bat flying past her roost. His regular route took him near the high bough of the pine where she perched. He always disappeared when he left the pine, but the Grouse found him again as he returned through the top of an oak tree. Drowsily she would wait for him to come into sight, until finally he seemed only a rippling of the dusk itself. Then night moved over the earth and she let herself lose him, and everything, completely.

The bat was one of the animals that occupied the Grouse's neighborhood while she slept. She knew there were others, some less harmless. Daytime creatures like the Grouse possessed their world only half the time. When they rested, a new population came out to live in it, and change it. Those waking in the morning could not assume that they were safe because they had seemed safe when they went to sleep. Before any one of them stirred, he was ready with his defense—ready to leap, run, fly, pounce, or bite.

The Grouse's defense was stillness. After she opened her eyes on the morning of June eighteenth, she made no other sign that she was awake. A weasel might have been touring the tree on a search for roosting birds. Or an owl might even then have been focussing its eyes on the Grouse's blunt-shaped form. She would not turn her head and help any predator to find her. Few of them could remain quiet as long as a grouse could. She would continue to seem but a dusky pocket between the bough and the trunk of the pine.

This habit of stillness allowed the Grouse to gather other impressions besides warnings. In perching patiently, perhaps she had time to hear that the wind's grave sound was blended of different kinds of murmurs from various sorts of trees, and to see that the wind folded the clouds as smoothly as water flows. She may have sensed something of these happenings, since no tension in her eyes suggested that the dawn was too slow.

But daylight brightened. As soon as the trees were definitely green, the Grouse began to stir. She eased one foot and then the other, loosening its night-long clasp on the bough. She relieved the muscles in her skin by fluffing her feathers and contracting them again. She stretched her wings. All these motions were inconspicuous, however—nearly as smooth as the motions of the clouds.

Now the Grouse tilted her head to see what was causing the sounds below. Bark had fallen to the ground. Claws were grasping the ridges of the trunk, not far up the tree but coming higher.

Since it was a climbing, not a winged, animal that approached, the Grouse might have escaped danger by flying into the sky. But it was a grouse's way to hide until hiding had become hopeless, never to fly until discovered by an enemy. She had no proof that the invader knew of her presence in the tree.

On a fir bough that extended into the pine, the Steller's Jay was perching. At the sound of the claws, his long beak opened in a rasping squawk. He flew down to investigate, and his excitement increased. As fast as he could work his beak, he screeched, partly harassing the enemy, partly warning other Beetle Rock animals to be on guard. A second jay arrived and the clamor doubled. Then three more came.

As the invader climbed, the jays flew up from branch to branch. By watching them, the Grouse knew how far the enemy had progressed. It was coming rapidly. Around the bough below appeared a large fur head. Wildcat! Now the Grouse can see its shoulders, too. Her toe and wing muscles tighten, but still she does not fly. And she was, after all, to be safe. For the cat had stood all that it could of the jays' siege. It paused on the pine trunk, and then started to back down. The jays followed. On the ground the cat streaked away among the boulders and escaped the birds. They flew back into the pine and fluttered and called, but the urgency had gone out of their voices. One by one they left the tree.

There was a practical reason for the grouse habit of freezing rather than flying when a predator was near. Grouse wings moved with a clatter that would attract attention instantly. No day could begin, however, without a flight from the roost. The bird walked to the end of the branch with the unobtrusive step that also was part of her defense. She raised the short round wings above her back and dived from the bough. She felt her body fall before the wings came down and began the flapping that checked the drop. She headed along the edge of the Rock, bound for the stream at the foot of the slope. As soon as she felt the canyon wind flowing under her breast, she rested upon it, and glided out over the Rock's rim and the trees below as smoothly as if she were one of the great flyers, a wild goose or an eagle, or as if she were expressing her true temperament even in flight.

The wind was letting her down, but not too soon. Directly beneath was the stream. The Grouse tilted her wings so that her body was nearly vertical, and depressed her tail and spread it fanwise. Then she fluttered, not very gracefully or surely, under the streamside branches and alighted at the edge of the water.

After the exertion of the flight she slipped into hiding among the roots of a pine, draped with dry sedges along the bank. There she groomed her feathers. She came out and walked along the sand, slowly. She took several sips of water, pausing after each one as though to take the sounds of the water, also, into her. The stream spread out in a small pool here. The silken sheet that fell into it formed bubbles where it struck. The Grouse watched them float down the side of the pool, whirl around at the lower end, and scatter. Finally all the distress had gone out of her, even out of her eyes.

Something else was in her eyes—a sort of expectancy. She moved downstream, past a stretch of water tossed into foam by the rocks, and came to another pool. There she found the companion she met at the stream each morning. He was the Mule Deer Buck, with whom she had one of those attachments sometimes formed between animals of different species.

The Deer was drinking. His mossy antlers lay almost in the ripples, but his ears were back to catch any sounds from the shore. One ear swung in the Grouse's direction. Perhaps he had heard even her light footfall. The Grouse saw the black tip of his tail flip in its white circle. She came down on the sand.

When the Deer's thirst was quenched he raised his head and stood gathering the scents and sounds about him. The Grouse at his feet took a beetle off a wet pebble, two robber flies from the sand, and a caddis fly from the water. Then she walked up the bank, climbing under the bright leaves of a hazel bush and through a clump of blue flowering lupine. The Deer made an easy leap to a ledge at antler-height above the stream. There the grass was finer than the sedges below. He began grazing on the tender tips.

The Grouse, too, had reached the ledge, where there was a meal of insects for her. The first scrape of her beak brought up a white beetle grub, interrupted at his own meal on a root. She found a cutworm, slow in burrowing down for the day, and a hard-shelled wireworm. She picked several ants off the ground with nicely directed snaps.

The Buck let his grazing draw him near. A small murmured note arose in the Grouse's throat. It sounded as if her quality of quietness had become, instead, gentleness. The Deer had no voice with which to answer, but his ears flicked each time that he heard her.

Now what had startled him? His head was up and his ears reached with a pointed quiver for a threat that the Grouse had not caught. She watched him closely. At all times she was aware of his ears; as long as they waved loosely she could risk a quick conspicuous dart for an insect, but when the ears turned together attentively, she made herself motionless.

This time she froze, under a seedling fir. Was the Buck hearing, perhaps, the scoop of a paw in the water, a scratching for fleas, or even so slight a sound as another animal sniffing a scent? It was, in fact, no sound that had alarmed him, but the ceasing of one. Two gray squirrels had been pursuing each other over some fallen oak leaves farther up the bank. Suddenly the quick tapping of their feet had stopped. Had the squirrels fled from an enemy? The chain of caution in the forest linked together many creatures. It had reached the Deer, and through him even the Grouse.

When the Buck could not sense any reason for the squirrels' abrupt silence, his ears swung away from the bank and his mouth went again into the grass. The Grouse took a step and pecked up an ant which had been running back and forth below her eye while she was still. The chain of caution had not extended to that small creature.

But before the Deer had nipped off more than a mouthful, his nose was flung again into the air. Now his whole body was alert, for he had caught a dangerous scent. He moved along the ledge, his gait tense and distorted. Then he was away, heading for the upstream slope. Soon even the beat of his hoofs had vanished.

The Grouse did not know what the danger might be; only that it was downstream, since the Buck had gone up. Before he had taken the first leap she had slipped under a fern. There the pale marks on her shadow-colored plumage matched the spaces between the leaves, larger near the ground, small on her head among the finer tips of the fronds. Her root-shaped feet branched where they stood at the base of the plant.

Before the warning, a junco, a fox sparrow, and a purple finch had been singing, sending out their voices boldly in order to be heard above the stream. Now their songs, and all other living sounds, were gone. The animal that was coming must move about in a silent forest, in a silence of its own making.

Up the bank slanted an animal path, marked with tracks of bears and coyotes, dappled over with pits formed by smaller feet. The Grouse held her eyes on this trail. Now a part of the spotted earth seemed to be sliding. No—there was something upon the path, spotted too. It was the wildcat.

The trail passed within the length of a cat's spring from the Grouse's fern. The cat's powers of smell might not reach that far, but her vision certainly could. This was a true test of the Grouse's defense. No jays were here; her safety depended upon her coloring and her skill in freezing. She seemed as motionless as the brown earth, with no sign that she breathed, no pulse of breast feathers to show a heart-beat. She was alert, ready to fly as a last resort, yet she looked as if she had removed herself already in some inner way, as if even her emotions were helping to disguise her by avoiding any intensity that might attract the cat's attention.

As the cat drew near, her eyes were on the water. Perhaps she was thinking of fish. And then she was down at the end of the path, slipping over the streamside boulders and onto the sand. Her belly low, her shoulders strong beneath her fur, she crouched and drank. Smoothness made the motions of the cat almost invisible. It was much like the Grouse's smoothness, but the Grouse's was for escape, and the cat's for capture.

Now the wildcat had flashed from the water and pounced at the edge of a rock. Did she think she would catch a mouse? She should have known better. This shrew under her paw had been squeaking to announce that it was a shrew, relished by no animal except an owl. The cat sniffed at its body, even smaller than a mouse's, and drew away.

The cat sat on the sand and watched the patterns of spray, existing only as something remembered. Then the Grouse allowed the lids to close over her eyes. For no cat was there. She had gone on, with her scarcely perceptible tread, to surprise other creatures. The Grouse remained frozen until she heard the lyrical song of the fox sparrow again.

The Grouse left the stream and started up the slope to Beetle Rock. Whenever she could, she remained against backgrounds colored like her feathers. She climbed near an animal trail, but went along at the side, under grasses and brush and seedling trees. The branches over the animal trail were high enough to clear the shoulders of the bears, and the Grouse felt exposed under them. Her own trail was as secret as a nest.

Even when she was walking, it seemed that she never quite emerged from stillness. She never risked discovering too late that she should not have taken a step. Before she had gone far she saw a red and black millipede lying in a spot of sunlight. A thrust of her beak would have reached him, but she moved forward until she was above him and could take him with a less conspicuous peck.

She strolled on; but suddenly waited, for something had flashed orange above the top of a rock. There it was again—a chipmunk's tail. It might have been the twitching ear of a coyote. Approaching a chinquapin bush, she saw the leaves on one branch shaking faster than the leaves stirred by the breeze. She froze while she watched them. A pair of juncos flew out. A flicker glided from a pine tree to a fir; its shadow passing over the ground checked the Grouse with a step half taken. Why did the jay screech? Was the puff of dust raised by the breeze or an animal? What had started the rolling pebble? The Grouse stopped until she knew. Since she placed each foot directly in front of the other, in the center line of her body, her weight was always perfectly balanced and she could become motionless the instant she sensed possible danger.

The flowing quality of her motions was all the more remarkable because the Grouse had an injured leg that summer. When she walked more quickly, she limped. Early in May she had been caught by the Coyote. She had been sunning herself, with her mottled feathers outspread on the granite gravel, when the Coyote had appeared from behind some brush. He had not smelled her, for he was following another scent, but the Grouse was directly in his path. Seeing her, he had sprung for her. She had tried to fly, but his jaws had closed over her leg. Beating a wing in his eyes, she had forced him to let her go, then with an extreme effort she had reached a low branch of a cedar.

At first she could not stand on the leg. She got over the ground by fluttering, and drank from melting snow near Beetle Rock. But within a few weeks she had returned to her former habits, including the daily trip to the stream. The leg was sore and stiff and the Grouse instinctively treated it by using it.

These recent weeks had been lonely, for it was the nest-building season, when the covey was broken up and the pairs of birds stayed in deepest hiding. Late in April they had flown down from the higher mountains where they spent the winter under the snow tents of the western white pines. Almost at once, then, they had found their mates of the year, and had gone down into the thickets, to make their wonderfully disguised nests on the ground, to fill them with eggs and remain secluded until the eggs were hatched. But now it was almost time for the covey to assemble again. First the cocks would appear, leaving the later training of the chicks to the hens. And the families would be out in the brush. The Grouse found one of the hens with young chicks as she climbed the slope on June eighteenth.

The Grouse had fluttered to the top of a log for some termites. On the other side she discovered the hen with her fledglings. The speckled balls trailed about their mother as closely as if they were attached to the ends of her feathers. Down the shadow of the log moved the family. The Grouse watched them go, then turned to a flat granite rock, circled by brush, where she usually stopped to rest her leg.

The mid-morning sun had made the rock comforting to a bird's flesh. The Grouse let herself down upon it and soon felt the heat beginning to come through her feathers. She lay tilted, with her injured leg in the sun, and expanded her tail and wings so that the heat could sink into the muscles and spread the oil that lubricated the feathers.

Enclosed by the brush, the Grouse was hidden from any animal passing on foot, but she could be seen from the branches of surrounding trees. She had relaxed in the warmth, and her eyes were partly closed, when, starting slowly, the air began to vibrate with a rhythm that was half sound, half pulsation. There were six deep muffled beats, followed by silence—silence except for a grasshopper's brittle click. The Grouse listened to the hopper, waiting to hear the low, slow tones again. A chickadee called and a robin chirped. Perhaps the Grouse only had dreamed the disturbing rhythm. No, it had begun again. It was beating for her, as she knew. A male grouse in a ponderosa pine had seen her and was calling to her.

This probably was the wildest of all forest voices. Its resonance did not even come from the male bird's throat, but from his flesh, tightened over expanded air sacs. The sound was so intangible that one must listen with nerves as well as ears to hear it. Each cadence of beats was like a thought, a wish, expressed only for the Grouse. The cock even used ventriloquism to hide his whereabouts from most listeners.

The Grouse seemed no longer at peace on her sunny rock. She moved out from the brush, ran a few steps to gain speed, then rose on her whirring wings and flew to a sugar pine. The cock could see her there, too, and continued his calling. She walked to the end of a branch, came back, and settled restlessly near the trunk. She preened the feathers of a wing. She fluttered to a higher branch and turned so that she faced the gorge instead of the slope.

Now instead of the gullies she had passed, the Grouse saw canyons cutting mountain ranges; instead of boulders, granite cliffs; instead of leaves above her, thunderheads with blue shadows. A world that expanded infinitely lay around her, and in the calls another world that might expand.

She flew again, passing the male's tree. She could see him clearly. He was the finest cock in her covey. Weeks earlier, he had fought with another male, hoping to please the hen that the Grouse had met with the chicks. He had won the fight, too, but the hen had chosen the loser, as female grouse so often curiously do. The most impressive cock in the grouse community was single this year, as was the hen with the injured leg.

But now her injury was almost well. The cock must have realized that it was, for he was putting all his talent into the serenade. As the Grouse approached, he lowered his head, inflated the sides of his throat, and with an effort that seemed to take his entire strength, forced out the air in the rhythmic beats. Before the last short beat, the Grouse had alighted in an oak tree.

Although she was farther away now, the cock continued to call. But a new sound conflicted with his message. Somewhere up the slope a human being was chopping wood. The sound was deliberate and steady, plainly driven by a mental will, not the free wish of an instinct. The strokes of the ax broke through the bird's wild magic. In any case, could the cock's hope be fulfilled with the year so far along? If the Grouse responded now, would she not come to a day when she would be mothering a late brood of chicks, too young for the autumn migration? The problem could hardly have been so definite to her, but her instincts, or some inarticulate part of her nature, would weigh it and give her the right reaction.

DIRECTLY ABOVE THE TREE that the Grouse now had reached stood Beetle Rock. To get up over the lofty wall would have been difficult for her, on either feet or wings, but along the north side of the Rock was the draw that now sucked up some of the canyon breeze. The Grouse launched herself into this current and let it carry her to the Rock's upper edge.

She alighted in the fir that would be her headquarters for the rest of the day. Beneath its branches was a sheltered lookout where various animals gathered to rest, to enjoy the filtered sunlight, to take dust baths, and to watch the doings on the Rock. The Grouse's friend, the Deer, came there to chew his cuds. A golden-mantled squirrel had its burrow among the roots of the fir, and pairs of juncos, chipmunks, and lizards lived near. They formed a little neighborhood in the larger Beetle Rock community.

This day the Grouse stayed for a while in the branches, for rain had started to fall as she was flying from the slope, and her weather sense told her that there would be more of it. She crouched on a bough close to the trunk and drew her head into her shoulders. Sheets of water soon were blowing over the Rock, lightning was streaking above it and thunder was hurling crashes upon it, but the Grouse was as snug on her perch as if she were in a feather cocoon.

As soon as the storm was on its way down the mountain, she prepared to come out of the tree. The ground near the trunk had remained dry, but at the edge of the branches' canopy, enough moisture had fallen to bring appetizing bugs and other insects to the surface. The Grouse tilted forward and started to raise her wings for the drop. She looked down to choose the place for landing—and saw an immense fur back beneath her. Her toes tightened to hold her on the bough, and habit kept her quiet. But her eyes, wide and hard, revealed that terror had possessed her. The animal she saw was the Coyote who had caught her leg. He could not reach her in the tree, but the unexpected sight of him was a shock.

He was not aware of the Grouse. He had trotted out of the thicket and passed under the fir on his way towards the draw. The Grouse saw him sniff around the manzanita and wild lilac bushes, pausing from time to time with his eyes on the burrow holes of squirrels and chipmunks. Suddenly, as if he had caught a clear scent, he bounded under an oak tree, where the family of weasels had their home. From a distance the Grouse could see him digging to the nest, perhaps to the young, for now the cries of the mother were shrill.

As the Grouse had been watching the Coyote, the Mule Deer Buck had come. He stood under the fir, also with his eyes on the Coyote, and fairly quivering with alarm.

A ground squirrel was whistling its warning; at a hole in the fir trunk appeared the face of a wakened flying squirrel; the Deer Mouse ran up the tree in a panic and took refuge with the squirrel. Fear circled out from the Coyote, spread from one little life to another so far that many trembled without knowing the danger.

What the Coyote found in the burrow could not be seen from the fir, but the screams of the mother Weasel expressed a desperate grief. In many such incidents she had played the Coyote's part, yet she was able to sound more outraged now than most gentler animals could have done. The frightfulness could not last indefinitely. Finally the Coyote loped down off the Rock. The Weasel was silent; the other animals abandoned their defenses.

But the Coyote left a tension that broke out in many small conflicts. The first to show it was the Buck. A nervous yearling doe had come beneath the tree. The Buck was browsing on twig-ends of the lower branches, trying not to let them touch his tender growing antlers. The doe kept getting in his way, prancing, whirling her ears at imagined threats. The nearest danger, the Buck's annoyance, she did not sense.

The Grouse did not react with stillness to any disturbance in which she was not involved. She had come down from the tree and now got away quickly, on toes and wings, feathers bristling and tail upspread. She went down to a hollow in the Rock where pollen cones from a Jeffrey pine had blown. These were her favorite food; she would quiet herself by picking them up, one by one.

Soon the doe raced by with blood on her shoulder. The Grouse stopped eating and watched her disappear, then turned back to the pollen. But she found herself in another skirmish. The Lizard had trespassed on the hollow. Another lizard, owner of the hollow, must drive him out. As the two flashed and darted around the Grouse, she fled again, circling into the trees behind the Rock.

When she returned to the fir lookout she found the tempers at a climax. The Buck was walking around a coral king snake that lay coiled near the tree. Every watching animal was tense. The Buck's strained gait, and the menacing thrust of his head, showed that his fury could not be controlled much longer. Yet he continued to move around the snake, and the tension increased. Finally, with a swift leap in the air, he crashed upon his victim, his sharp hoofs together. Again and again he sprang upon it, until his strength was gone. Then he came back under the fir.

This time the Grouse had stayed. Now, choosing a spot where the tree's broken shadows resembled her markings, she folded her feet and lay upon the fallen needles. The breeze still blew from the draw, but she partly raised her feathers, allowing the downy after-shafts to fluff out and enclose her in warmth. The anger visibly went out of the Buck. A cud rolled up the front of his throat. He chewed it, at first with frequent pauses to concentrate on suspicious scents or sounds, then more steadily. Finally he knelt with one foreleg and the other, and lay down.

Many of the animals were sleeping, though not all. Around the Grouse and the Buck moved the golden-mantled squirrel, walking on its hind legs to reach for seeds in scattered heads of grass. From a cherry bush nearby, the junco tinkled a tiny gold song, and a mountain swallowtail fluttered by, bent on its delicate business. The Buck watched the happenings around him, gravely, but the Grouse seemed to turn to shy impressions of her own. In her eyes was an expression, liquid and remote, as if she were hearing bird or insect music pitched too high for many ears, or perhaps, most likely, she merely was feeling an exquisite inner balance related to the balance in her motions.

But the chance to be serene soon ended. The Grouse's very gentleness sometimes tempted animals to tease her. Towards mid-afternoon she felt inclined to take a dust bath. She pushed herself into the sun-warmed crumble of bark and needles. With her beak she threw some of the dust into the down of her breast, she rubbed the feathers of her neck and head into it, and with her feet she tossed it upon her back and under each wing. When she would shake it out, she would leave enough dust at the base of her feathers to smother the mites on her skin.

While her plumage was still ruffled out, the Grouse heard the Chickaree's squeals in the fir. The next instant he was down the tree and bounding around her. His rushes were a bewildering whirl of flying tail and pattering feet, and he barked, whistled, and growled at her.

The Grouse did not give him the satisfaction of a glance. Now she repeated the entire process of dusting, deliberately. Her unconcern infuriated the squirrel. He was coming in under her very beak; it really seemed likely that he would nip at her. She waited until he was racing towards her once more, then threw herself into a fine flutter. Dust flew all over him. She spread her tail, jumped up and down and hissed a terrible, if indefinite, threat. Part of the splendid display of rage was wasted, however, for the Chickaree already had fled up the tree.

The Grouse had begun to compose herself when the cry of a hawk tore over the Rock. Before it had ended, the Grouse was under the cherry bush, as quiet as if the cry had struck out her life.

The Chickaree no longer seemed important. A hawk was danger in its most desperate form. Even if the Grouse should get off quickly on her wings, she could not be sure of escaping another bird. The grip of a wildcat would have had at least the numbing quality of strangeness; most to be dreaded was the threat of a bird's claws and beak, like the Grouse's own but with sharp destruction flowing out of them. Again, from the tree next to the roosting pine, came the hawk's threat. Doubtless its eyes were scanning every break in the Rock, every bough of the trees, every sheltering bush. Could it miss the Grouse, so imperfectly hidden in the thin shade of the cherry? Once more the pitiless cry penetrated all the hiding places, as if aiding the search.

No other sounds told what became of the enemy. Time simply passed. While the Grouse stood motionless, the shadow of a branch on the ground moved from the right of her to the left. A little cloud that had been above the western ridge vanished below it. The petals of a lily partly closed. Even after the chipmunks were chasing each other again, and a pair of olive-sided flycatchers were bickering around their high nest, the Grouse remained still. But finally her eyes shifted to the Buck, who was standing now. As if his size and strength gave her confidence, she stepped under a pine seedling and began clipping off the needle tips.

But the Buck soon left. Another companion of his, a younger deer, appeared from among the trees, and the two went away to browse in the oak thickets. The Grouse strolled to a rain pool where a robin and the Jay were battling to decide which should drink first. The Grouse was a more forceful bird than either. She did not fight for her place at the water; she stepped forward and took it.

The day seemed to be coming to a thin and scattered end. The sun had lost most of its warmth and had not yet grown warm in color. The storm had stiffened the Grouse's leg. The cock had not followed her to the Rock, and most of her neighbors had disappeared. She was hungry; at least she could do something about that. She went back to the pollen cones on the Rock. It was important to start the night with a crop full of nourishment; here she could find it with little walking.

Looking up, she saw the Buck browsing on the brush in a granite gully. And gradually other animals came out to seek the insects, seeds, leaves, or whatever food would sustain them until morning. A sociable sort of activity was beginning again, when the Grouse noticed that the eyes of the Deer were tense with surprise. She looked to see what had startled him. It was the cock, spectacular in his courtship display, assumed for her. To have him appear so abruptly startled the Grouse, too. The feathers on her head rose nervously, and she took a few quick steps towards the trees.

Perhaps the Buck had not even recognized the cock. Ordinarily he was a rather formless bird, darker and with fewer markings than a female grouse. Now all his outlines had become sharp, and his soot-colored feathers had parted to show a brilliant pattern of black and white and flame. On each side of his throat swelled the sac of vivid yellow skin, surrounded with a flat wreath of black and white feathers. Above his eyes had risen orange crests. His tail was a light-bordered fan spread above his back, its reverse side a pom-pom of black and white down. And the bird's manner, too, was striking. He advanced with a slow, bowing step, dragging his wings on the ground and dipping his head from side to side.

Other creatures besides the Deer had paused to watch the cock. The Grouse led him away to a small clearing in the forest behind the roosting pine. The clearing was familiar to all the covey, but on this evening it was like a new place, for it was in a stage of lovely flowering. A mist of tiny pink gilia blossoms floated just above the ground, and bright little birds, chickadees, kinglets, and hummers flowed around the twig-ends of the trees. But the most exquisite transformation had been made by the tussock-moth caterpillars. After hatching in the boughs of the firs, they had let themselves down to earth on long silken threads, which still hung in the air, glistening in the late sun's light.

Back and forth across the clearing walked the two birds, sometimes under the open sky, sometimes in and out of the bordering shadows. The Grouse was a few steps ahead, not leading now, but delicately fleeing. Whenever the cock paused, she would wait, and both would fall into their sensitive stillness. But she could always perceive the movements of the cock as he started forward again, and she would slip farther away, as if his emotion were a wave that pushed her out of his reach.

Occasionally the cock beat one note of his call; more often he urged her in a low tone, hoarse and compelling. Each time she heard it she lengthened the space between them. She must avoid anything impetuous until her own wish spoke more definitely. If she had had the most analytical human intelligence, it could not have told when the autumn storms would begin—whether a nestful of eggs could safely be laid after June eighteenth. But the schedule that intelligence could not have guessed, the more elementary impulse of the Grouse would know. She was feeling for its decision while the cock was pleading. He seemed to insist that the question be settled now. And yet she was not sure. Once more she started to the other side of the clearing. On the way she realized that he no longer called her, and she looked around. He was not there.

Soon she heard the whirring of his wings. He might have gone up into the covey's roosting pine, which lately she had been occupying alone...but the whirring faded slowly; his flight was a longer one. Soon the Grouse discovered where he was: in a tree at the rim of the Rock. He had begun the rhythmical beating for her. He might yet return to the pine. It was the custom of all the grouse to go into the roosting tree in a ceremonious way. They first walked to the top of a massive boulder and from there fluttered to the lowest branch of a cedar. They rose through this tree, branch by branch, until they were level with a long bough of their pine. Finally they flew across the space between and moved in a procession to the top. No doubt the cock was watching to see the Grouse begin that ascent. It was not unlikely that he would fly back over the Rock later, and join her on the roost.

The sun was nearly ready to set, and dusk had gathered in the clearing where the Grouse stood motionless among the little flowers. Moths, like ghosts of butterflies, were beginning to cross from tree to tree. The Grouse's wings came together above her back, lowered with the lightest possible stroke, and lifted her to a low bough, not of the roosting pine, but of a fir behind the clearing. She moved in along the branch towards the trunk. With one soft flutter she alighted on the next higher bough. By walking partway to the end, she came close to the branch above. Feeling her way, with never an awkward step, she climbed, her pace becoming slower. Near the top it was timed with the sun. As the sun only apparently moved when its edge was observed against something still, so the breast of the Grouse seemed stationary until it passed a twig. At last she reached the highest branch that she could clasp securely. She balanced her weight and let herself down on the bough. The clusters of needles hid her, but through them she could look out and see the narrow glow on one side of the sky, and the stars on the other side.

The cock no longer sent her his thought. The only voice that reached her was the wind's, which was not calling to her nor to anyone. The integrity of her wildness had required her to make the lonely choice, but it would prevent her from dramatizing the sacrifice. If the cock's wishes and hers could not be fulfilled until next year, at least the wishes soon would be silenced—as soon as the covey would come together for their pleasant fall wanderings. Meanwhile, she and the cock were birds to whom waiting was not a vacant time, but a way of full living.

Even on this night in the strange tree, the Grouse was not entirely alone. Before she closed her eyes she saw the bat that she had watched from the pine roost. He passed her once, then again, and again. He was not a friend but he was someone comfortably familiar.

WHAT HAPPENED TO THE CHICKAREE

THE INSTANT THE CHICKAREE STIRRED from his sleep he was tense—he must leap at that jay! The jay has a long, sharp beak. The Chickaree has no weapons except his teeth, shaped to bite nuts, not bones. But even the defenseless can fight. The jay wants the Chickaree's mushrooms. He shall not have them!

The waking squirrel opened his eyes, and found himself in deep night. No jay was there, only jay feathers. Under the Chickaree was a whole bed of the feathers. As he slept, every breath had drawn in the disturbing jay scent.

Never before had he smelled jay feathers when he awakened. His nose quivered, searching for the odor of other squirrels, for the fur, musk, and pitchy paws of his family. Their scent had been strong in the hollow pine stub where he had slept during his ten months of life. Now he could find no trace of it.

The dead pine had not swayed; this bed was swinging. And the night noises were louder here than they had been in the stub. From below in the canyon came the cold, continuous roar of the river. A bear snorted as it stretched against a cedar and tore at the bark. Somewhere near, the jaws of a bat snapped over a moth. More familiar was the wind, hanging in loops of sound that lifted and fell as they were combed by the long pine needles.

Many times the Chickaree had aroused in darkness, when his mother had untangled herself from the sleeping young to leave the nest. He never had followed her then, but now he was out in the night, lost among strange sounds and scents. A faint light was coming over his shoulder. He turned to it, and in moving clipped off his drowsiness. Now he knew where he was—this was his new nest! This was the first nest of his own, the one he had been making for many days.

Only yesterday the Chickaree had finished the bed of the nest. Its base was soft twig-ends, cut from the branches, then collected on the ground and carried up the trunk. Twigs in place, the Chickaree had gone below to hunt for feathers. He had found a heap of them under a seedling pine—all Steller's jay feathers. The jay's captor, who had left them, had left no trace of himself. The Chickaree had sped up and down his tree, taking the feathers to his nest before any other squirrel might discover the treasure.

The bed had been finished by sundown. When a flap was made for the entrance, the entire nest would be complete. The Chickaree had not intended to occupy it until the hole was covered, but when he had placed the last feather, he had been very tired. He had curled up on the new bed instead of returning to the hollow stub. Sleep had descended swiftly as a hawk.

After he woke at midnight, the Chickaree lay within the dome of the nest, tail over his back, chin between his forepaws. He could look out into the dark trees stirring, and the sky. The moon was hidden behind the mountains but it brightened the clouds, many shining strands of clouds flying westward over the treetops.

The Chickaree crept from the nest. Softly, without a click of his paws, he ran out on a branch. Between the twigs he had wedged a supply of mushrooms to dry, to be stored later in knotholes. He would eat one mushroom tonight. Taking a fresh one, he went back to the crotch of the limb. He sat up with his back to the nest, turning the mushroom in his forepaws and nibbling the edge.

The sprays of the fir were clustered around him, a cover in which to hide and feel secure. But suddenly a cold call flooded through them. A great horned owl, in this very tree, had heard some animal move and would try to frighten it into revealing its whereabouts. The Chickaree had a panicky impulse to dart into his nest, but resisted it. He lowered his eyelids to conceal the shine. Then he froze, with the mushroom in front of his mouth.

Again came the owl's hollow call, and everything firm slipped away, as if the Chickaree had jumped for a branch and missed. He waited, knowing that the owl was somewhere in the black shade behind him. Now even the wind was still...though the safe white clouds raced as steadily over the sky.

Finally, after a long silence, the Chickaree heard feet strike the earth below. There was a tearing of roots and fibres, a tiny shriek, and quiet. The Chickaree remained motionless until he saw the shadowy owl fly across the Rock's granite field and out over the canyon. Then he began again to take bites of his mushroom. When it was gone he went inside. He spread his tail over himself, and under it was warm. But the owl had increased his uneasiness about the entrance. As long as it was open he would not be able to sleep soundly in his new nest.

He was roused in the morning by the voice of the Steller's Jay. Plainly the bird was excited. Danger to a jay often meant danger to the Chickaree, but he tingled with curiosity. From the nest hole he peered cautiously down the fir trunk and out through the boughs. When he found no invader in his own tree, he bounded to the end of the nest branch.

Other jays had begun to screech, and the squirrel joined their clamor. His cries were much like theirs, but he seemed to make them differently. He jerked, as if he were a little bag filled to bursting with bright sound that piped out whenever the bag was jostled.

It was a fine vivacious start for the day, to swing on the end of a bough making that shrill commotion. But the Chickaree soon lost interest. He couldn't discover the danger. Probably it was on the ground, a long way down. Later he would go below for something to eat, but he could begin with his mushrooms. Taking one of the wide, flat kind from its notch, he went back into the nest with it.

The morning was only a shade lighter than the night, but it was something definitely there, a dark brightness in the branches that sharpened the shape of the needles as water seems to sharpen whatever it wets.

Generally the Chickaree sensed the approach of day by the birds' doings. Today something had stirred the jays earlier than usual, but soon they quieted down, and chirps, as detached as waking thoughts, were coming from many small throats and putting time back in its proper step.

The Grouse broke from her pine and came sweeping downhill. For an instant she was in the Chickaree's tree, and then was out in clear sky. The rush of her wings fell quickly into silence. One rough cry filled the air as a raven passed, flying ahead of the flock that soon would set out down the canyon. A pileated woodpecker tapped a few times slowly, and stopped. And an olive-sided flycatcher called. All day its voice would plead, as if it never would lose hope of getting an answer. These sounds were only fragments, not yet gathered into anything sure, until a robin began to sing. It was the robin who had a nest on a lower branch of the Chickaree's fir. Its voice rose firm and clear through all the tree.

The Chickaree came out of his nest and headed down the steep trunk. He timed his steps exactly with sweet, liquid barks. First he went forward one set of steps and one bark, then a run and a medley of sprightly sounds to accompany it. Another slow step and slow bark, and the quick ones again; all the way to the ground he matched his gait with a kind of musical announcement.

The ground was always a foreign place, and the Chickaree paused before he went onto it. His hind legs were on the trunk of the tree, his bark-colored fur was against bark, and the push in his forelegs was ready to whirl him back if he saw or smelled an enemy. But he didn't, so he leapt into hiding below some seedling pines. He was starting to his old home grounds around the family stub, and to get there quickly was going partway through the brush.

To run would have been too dull a way to travel. The Chickaree sprang along in bounds, touching on front feet, then more than his length forward on back feet, bouncing over twigs and rocks softly and lightly. On each tree trunk he jumped sidewise, stopping to look ahead and be sure he wouldn't meet any enemy before he reached the next tree.

All the Chickaree's industrious energy turned this morning towards his wish to finish his nest. But first he must fill himself up with good nourishing food; no animal disregarded that need. As he passed, he saw juncos pecking among the grasses, trying to make their discoveries catch up with their appetites. The Chickaree galloped up a fir tree and out onto a branch. From its end a western wood pewee circled into the air to catch insects. The Chickaree leapt to a pine, bounding past a white-headed woodpecker drilling for beetles. These birds, like most of his neighbors, depended on chance to satisfy their hunger, and often were disappointed. The Chickaree never had known the insecurity of an empty stomach.

He and his family had worked together during the previous fall, gathering and storing nuts and berries. In the spring the young squirrels had scattered, but those near enough still came back to share what remained of the winter stock. The Chickaree went first to the draw where the cones were buried. At first there had been more than fifty of the cone caches, with up to twenty cones in each. On June eighteenth he found a whole untouched cache. These were only the small sequoia cones, but he took one in his mouth and jumped to the top of a log with it.

As he chewed through the scales to reach the seeds, he noted the happenings of the night in this familiar territory. A wildcat had left her scent in the draw and it made the Chickaree sharply aware of all the openings between leaves and tree trunks. A bear had passed; brown hairs caught on the spiny twigs of a manzanita told that the bear had pushed through to pull out the last of his woolly winter coat. A new chipmunk was stretched upon a boulder. He must have come since yesterday. He watched enviously as the Chickaree's teeth dived into the cone. Towards the tip the seeds were scarce, and the Chickaree lost patience. He tossed the cone to the ground and the chipmunk ran to it shyly.

The Chickaree saw that the nest of some mice had been torn apart. Who had done that? He dropped from the log and sniffed over the scattered fibres with a murmur, low and plaintive, as if he felt it a sad thing to be mystified. Near the end of the log were tracks of the Deer Mouse, and a little pit in the soil. The Chickaree remembered that he had buried an acorn there. He scooped the pit deeper, found that the Mouse had taken the nut, and turned away with a small, fiery mumbling.

The Chickaree then dug to another acorn, but one whiff proved that it had been wormy. When he buried it he had not been as keen as he was now in detecting the worm smell. The third acorn was a good one, with his lick sign still upon it. Spring rains had left a mat of leaves over the spot since he'd pawed out the hole and thrust the nut in deep with his nose, since he gave it the lick that meant "this acorn is mine." But he knew where it should be, and scratched away the leaves and found it.

Finally he unburied a Jeffrey pine cone. It was a big one, half as long as he and weighing a third as much. He couldn't bound through the boughs with so heavy a cone, so he carried it back to his fir by the ground route. With the cone in his mouth he started up the trunk of his tree. Once he had to stop, cling to the bark with his hind feet, and shift the cone with his forepaws so that his teeth had a better grip. But soon he was past his nest. He went on until he had reached the highest crotch where he could brace himself.

This was his favorite perch and his favorite kind of cone but he worked as if he had only one wish—to be through and away. He held the cone by its ends, turning it on the branch, pulling out the tight scales with his teeth. The two seeds under each scale he stored in his cheek until he had a mouthful to chew.

While the Chickaree's cone scales flew, the sun rose over the sharp white skyline, high on the east. Soon it was round. Its flame was licking the edges of the mountain peaks, and a violet radiance poured down the snows.

The world became crested with light. Above the shadowed ranges, brightness touched granite pinnacles and domes. It washed along the tops of ridges, now the ridge beyond the canyon. It reached the trees at Beetle Rock. The delicate fire was leaping from one green crown to another. It was on the Chickaree's tree—on his cone and the glinting hairs of his paws. It was making his cream-white belly rosy.

Around the squirrel, in the sky, halos circled the Jeffrey pine tassels as sunlight gleamed on the needle tips. The darker sprays of the firs were sparkling; the plumes of the sugar pines glistened more softly. Green needles were clustered with needles of light, green shine was reflected on shadows, green shadows lay behind all the shimmering. Nothing was still, nothing was solid or firm in this high field of the treetops, but the field belonged only to airy creatures, to the birds and the few squirrels that broke up into it.

The Chickaree looked up from his cone.

"Qui-ro!" he cried.

The whistled call shrilled out over the trees, down into the canyon, up towards the snows.

"Qui-ro! Qui-ro!" Then faster: "Qui...qui...qui...qui... qui, qui, qui, qui, qui...qui-ro!"

The Chickaree gave the cone core a wide throw over his shoulder. His paws and face were sticky with pitch. He chewed the pitch off his paws, then rubbed them over his nose, scraped his chin on the branch, and gave his fur a licking and preening. Now he was fed and clean and ready for the work of the day, the weaving of the flap for the nest hole.

The work was urgent, because the Chickaree felt exposed without the hanging, and because nest-building belonged to the spring, long since past. This actually was the second nest he had made. The first he had started on time.

During the early warmth, when snow was thudding down from the branches, he had felt a vague restlessness in the hollow stub and his mother's nest. At that time there had come to Beetle Rock a new little female, with gracefully daring ways. He had spent several lively days chasing her and fighting other males that would not stay away. Finally he'd caught her, so completely that he knew that game was over.

At once she had begun to line a hole in a pine with grasses, soft for hairless new-born squirrels. That nest would be the Chickaree's, too, on the coldest days of the following winter, but the inside of a tree didn't appeal to him for the summer. When his mate had begun her nest, he also had suddenly wanted one, but he built his own out on a cedar bough.

That first nest had been huge, with more and more twigs and fibres woven into it. He never used it. When it was nearly finished, lightning struck the tree, the bough crashed, and the fine new nest was scattered over the wet ground.

The Chickaree had begun another the next day. The top of a tall fir had been broken off by the wind, years before; from the break one limb grew out horizontally and another straight up. On the trunk, in the nook between the branches, was the second nest, woven more firmly than the first.

Now that only the hanging remained to be made, perhaps the nest could be completed before night.

The Chickaree got strips of hairy bark from his old cedar. Wonderfully the knowledge of how to make the curtain was in his paws. It should be attached to the top of the entrance but left loose below. He began by pushing the shreds of bark into the matted twigs over the hole. When he had drawn them through, he let the ends hang loose for weaving. The work would have been easier if he hadn't had stubs for thumbs, but he handled the fibres by folding his four fingers over against his palm. Usually he could make the shreds go where he wanted them. When he couldn't, he let impatience fly from all over him, from his tongue and tail, his shining brown eyes, and his erect little ears. A few times he was so annoyed that he stomped his feet, sputtering and whistling. He wove so steadily that the nest began to seem finished, though the hanging was still a paw's width from the floor.

For a chickaree, every season brought a new undertaking. After the nest-building of spring would come an exploration of his trees. The Chickaree had appropriated six, and they still were new to him. He didn't know all the possible leaps, the danger spots where boughs swung too loosely, and where lichen hid twists and breaks. Now, with the nest being completed so late, the Chickaree had a strong impulse to be gaining that knowledge.

Perhaps it was because of that impulse that he would finish the hanging with moss. He could get moss on the old family stub, circling over through his Jeffrey pine, sequoia, and sugar pine. He started by dropping lower in the fir. Fir boughs were so dense that in that tree he moved like a fish in a pond of water, with a flowing in quick, sinuous curves. An easy leap took him to the Jeffrey pine, where he galloped on branches long and firm in the wind. The sequoia boughs were twisted, but his claws gripped the fibrous bark so that he dared to run swiftly. Each kind of tree called for a different skill. The Chickaree hadn't yet the sure footing of an older squirrel, but he was learning to depend on the sensitive tail that helped to correct his balance, and the set of long tactile hairs on each arm that guided his paws.

Now the Chickaree would leap into his sugar pine—a wide space to cross, but he has run to the end of the sequoia bough, is in the air, outspread, as flat as he can make himself. But the breeze is blowing the pine branch! It is tossed far over to the side! He cannot reach it—he will fall! As he gropes frantically, instinct turns his flattened tail. It has twisted, changing his direction. He swerves to the branch, his feet touch it, and he clutches the downy foliage, frightened but safe.

For a moment he clung motionless, while the branch dipped. He looked down more than sixty times his own length to the dense brown earth. His heart still raced—but the amazing thing that his tail did was a new trick that belonged to him. It was something more added to his skill in the trees. Digger squirrels subsisted on foods that they could gather without courage, nimbleness, or grace. Whether the Chickaree was inspired by a more ambitious appetite, or some lift of the heart, he left the earth's safety for the birds' world, substituting daring and agility for wings. He never could sail into clear sky as the Grouse could, but the perilous leaps he made so willingly kept him in high, resilient spirits. He took the bright chance, and its touch fell upon all his ways.

Now his bounds as he crossed the sugar pine fairly glinted.

"Qui-ro!" he cried.

All the rest of the way, each of his motions tweaked out a joyous musical sound.

While the Chickaree had been weaving his flap, clouds had risen and covered the sun. Rain had begun to fall as he came through the trees. Now he would get the moss quickly and return to his fir by the shortest route.

He climbed the stub, passed the nest hole with his family's unforgettable scent, and ran up to the top. The rain was coming down heavily, so he pulled off his mouthful of moss at once. He turned—but stopped, horrified, to see the Weasel and three of her kits climbing the stub. One weasel was a frightful enemy. Now four were blocking his way.

The weasels had not seen him. They were going into his old family nest. They disappeared, one by one, while the frozen Chickaree watched from above. He slid back out of sight on the snag. There he crouched, legs crowded under him and his tail in a close curl over his back.

Soon he was the center of one of nature's wildest battles, a mountain storm. Like the snarl of a wolf, its thunder leapt. The stub under the Chickaree quivered as if it had been struck. Around him the trees and the wind began to struggle. The boughs bent, swung, defended themselves against the force that was beating them. Some broke. The end of one fell upon the Chickaree. It stunned him, and he rubbed his paws over his head to try to clear it.

Rain pounded him, plastered the fur to his small, chilled body. It seemed to fall on his naked skin. And lightning flashed with unearthly brilliance, revealing him to a great horned owl in the nearest tree. To his nose came a blend of wildcat, coyote, and weasel scents, the odor of creatures that live by taking lives, the transmuted scent of their victims. Twice the Chickaree started to risk a dash past the weasels, but each time he smelled them he moved back. The dangers of his treetop life had schooled him to keep his nerves in order, and that training was helping him now to resist panic.

Finally the great anger was discharged. The storm, like a wolf with a full belly, moved away. Its thunder was but a muttering in its throat; the lightning's gleam was mild. Even the owl did not look so threatening, seen more clearly as the shower thinned. Its head was hidden against the tree. Soon sunlight slipped from the clouds and fell upon the Chickaree benignly.

A scratching below meant that the weasels were leaving the hole. The Chickaree peered over the edge. One was emerging from the nest. It turned...down. Another came out and descended the trunk. Another and another. The Chickaree was safe. He shook the water from his fur, rushed back to his tree and up the trunk to the nest. He looked within. Undamaged by the storm, it was dry and snug.

He had forgotten to bring the moss, but in himself he could feel a drag, and the impulse to lose it was stronger than his urge to finish the entrance flap. He lay along a branch, hind legs outstretched, paws under his chin, tail loosely hanging. His consciousness sank, and rose, and sank. For a while he was lost, then his eyes partly opened and into them swam the image of lacing branches and shimmering needles. Near him a bee's gold body hung between the two blurs of its wings. An Audubon's warbler flew up through his tree, easy as an indrawn breath. Behind it a downy feather floated, and caught in a cobweb.

All disturbing scents from the ground were drowned in the warm pungency of cedar, fir, and pine needles. The trees were still. The only sound was the soft trumpet of a red-breasted nuthatch, lower in an adjoining tree. It piped one note unceasingly, a sweet monotony on which the Chickaree drifted back to sleep.

THE CHICKAREE WOKE, stretched, yawned and curled his tongue up over his lips, and bounded away for the moss. Speeding back, he reached the Jeffrey pine next to his fir—and stopped. Two bear cubs were climbing his tree.

The squirrel picked up one forepaw, and the other. Neck far outstretched, he stared at the unbelievable. The moss dropped from his mouth then, as he whirred and screamed. His little feet danced. His tail jerked in a frenzy.

The cubs continued to climb. Hooking claws into the bark, they moved up one foot at a time. Would they go as high as the nest? Would they trample and break it? The Chickaree drew himself up on his haunches, holding his forepaws out, limp, with anxiety.

He saw the higher, black cub turn out on the bough below his nest branch. The bear crawled along as securely as a giant furry bug. When he came to a fork he settled down among the needle sprays. Now the second cub, a brown one, reached the limb. Standing on it, he put his nose on the trunk of the tree. He had smelled a beetle working within, and tore off the bark to get it.

Again indignant cries burst from the Chickaree, seemingly all the angry sounds he knew, whistles, spits, sputters, and growls. There was nothing else he could do to frighten the bears. He knew he must not go into his tree, near those arms that crushed squirrels with one sweep. But that was *his* fir, *his* bark, *his* beetle—especially that was his new nest, just over the bears' heads. Would they scatter it, as his first nest was scattered?

Since the Chickaree could not fight the bears, he would fight another animal, any that he could find. In the society of the treetops, some creatures like a combat; others will give up food or a perch for peace. The Chickaree knew what to expect of each, but today anything that moved would receive his challenge.

The flip of a wing snapped his glance to the trunk of a pine where a red-shafted flicker had alighted to lick up ants. It was a quick jump over. In along a branch the Chickaree whipped. And now he was racing up the tree, all his feet off the trunk at every gallop. But the flicker would not defend its meal. It was flying away. The Chickaree followed to the end of a bough, too far behind even to startle the bird. He might have known that the flicker would not fight with him, but he stood on the branch, thumping in exasperation.

He turned back into the pine, scolding, flicking his tail. Then with a bound he picked up speed. He had remembered a western wood pewee in the adjoining fir. A pewee was a spirited bird. Its nest was on a dead limb at the top of the tree. The Chickaree darted up and out from the trunk. The pewee was not there.

More slowly the Chickaree moved along the branch towards the nest. He never had been able to come so close to it before. He drew up to the edge and looked inside, a forepaw folded against his breast. In the grass-lined cup lay three eggs, ivory speckled with russet. Still as pebbles they lay, but much more fascinating. The rigid branch was quiet, and no creature called.

The Chickaree looked at the eggs and something unexpected happened. His head bobbed forward. For the first time in his life he wanted a bird's egg in his mouth.

The three eggs might have been reduced to two, perhaps to none, but the mother returned. Now the fury was in another breast. The pewee flew at the Chickaree with a scream and a clapping of her beak. He did an instantaneous turn and raced back off the branch. He wanted only to lose that angry bird. Soon he did. He retreated through boughs so dense that the pewee could not follow.

The Chickaree ran to the Jeffrey pine and out for another view of the bears. Now the brown one was crouched on the branch. His brother had come in and wanted him to move so that he could get onto the trunk. The brown one looked off into the trees, pretending not to see him. A cuff from the black bear started some boxing, but soon both cubs quieted down. Apparently nothing else would happen for a while.

The Chickaree stomped and squealed. When his little tempest accomplished nothing, he set out on another hunt for trouble. Bounding through the trees, he passed a bat, limp in sleep between a curtain of lichen and a pine trunk; no fight there. He skirted a hole where a family of flying squirrels were nesting; no hope from them either, nor from a chipmunk "wheeking" from a lower bough. Then he saw that a great gray squirrel had dared to come into his sequoia.

The gray squirrel had reason to know that he would not be allowed in the Chickaree's range. He was twice as large, but the Chickaree was more than twice as fiery; he had proved that several times. For the gray squirrel to intrude today was a special outrage—and a comfort. Now the Chickaree could really let his temper fly.

His warning cries kept pace with his feet as he sped through the Jeffrey pine, leapt into the sequoia, and was down to the gray squirrel's branch. The gray squirrel was off at once, jumping into a sugar pine.

The Chickaree followed. Now he saw only a frost-colored tail, flowing, dodging, curving, streaming ahead of him. Down long branches he chased it, over the fir sprays, across gaps of clear space, up the tree trunks, out on dead snags. He was gaining on it. Though the tail had more balancing skill than the Chickaree's, the Chickaree's feet were faster.

From the topmost boughs to the ground and up again went the chase, into one tree after another, with a rattle of falling twigs, bark, and lichen. The gray one flew silently, but the Chickaree had enough energy to overtake the big squirrel and to squeal his resentment. He saw the gray squirrel leading towards a jump that a small squirrel could not make. He had only one branch on which to catch him or lose the chance. As the gray squirrel ran out on that branch, the Chickaree jumped to the bough above him, streaked to the end, dropped on the tail below, and caught it in his teeth. He bit off the tip and the gray squirrel screamed, then was sailing over the gap that no chickaree could cross.

The little squirrel crouched on the branch until he had caught his breath. But he was not spent. Returning through the trees to his own range, he bounded as though he alighted each time on the tail of an enemy. Now he must have more excitement.

He remembered that the Grouse and the Mule Deer Buck often rested under a fir at this time of day. He hurried out their way. Beneath the drooping branches, the Deer was lying down and the Grouse was dust-bathing. Like a shower of sparks the Chickaree was upon them.

The Grouse showed no sign that she was aware of him. Crouched in the crumbling needles, she was throwing footfuls of the debris under her wings and over her back. She wriggled so that the dust would sift down to her skin, and then tossed more of it over herself.

The Chickaree began his teasing of the bird. He leapt forward and back, on one side and the other, whipping his tail, squealing, whistling, and chirring. He would dart up, poke his impudent little face into the Grouse's face, and bound away. The Grouse continued her unhurried bathing, but into her eyes was coming a wilder gleam.

Her rising feathers made her suddenly huge then, as with a violent fluttering she threw her dust all over the Chickaree. Her tail stood out like a circle of weapons, and she hissed at the squirrel, and leapt up and down in a way truly terrifying. But he was gone, already back up the fir, pealing forth protests. He wanted the whole forest to hear how the Grouse had abused him.

He climbed to the top of the tree and immediately came down again to the ground and raced out past the Grouse. She could not frighten him. He was thirsty, and she would not stop him from getting a drink.

But the Chickaree had let himself, this one time, become too agitated. He ran out to a rain pool on the Rock that was too far from the trees that were his proper haven. He put his mouth into the water and began to lap. Then his tongue froze. For he heard the scream of a red-tailed hawk.

This was not the high scream from a hawk remote in the sky; on the Rock itself was this threat. Streaking back to the fir, the Chickaree scrambled up the trunk. He climbed to a refuge among the branches, and there lay while the shrill voice cut again through the air, and again. A long silence followed, until it seemed sure that the hawk must have flown away or found its victim.

The Chickaree had nothing left for this day. Back in the pine nearest his own tree, he stretched wearily on a bough and watched the bears. He did not even protest as the brown one began to climb. The cub was on the trunk, as high as the Chickaree's nest branch, and was going higher.

The Chickaree had not seen the cubs' mother, the Black Bear, but now she called from the ground. One cub answered with his hunger cry, "Muh!" but neither started down. The Bear summoned them again. The brown cub continued to climb and the other lay on his bough. The mother started up the trunk. At this new threat to his nest, the Chickaree pulled himself erect.

But the brown cub, hearing his mother's claws on the bark, was coming down. Now he was past the nest. Both cubs were following the Black Bear out of the tree. The Chickaree moved out on the pine bough. He waited only until the mother had reached the ground, then leapt across into his fir. He entered the nest. It was exactly as he had left it.

He lay inside. The hanging still was not finished. Under its shreds the Chickaree could see that the cones of the sugar pine were nearly ready to harvest. Those in his own tree, too, soon should be cut. Tomorrow, after he had completed the weaving, he would learn the location of every cone in his range, and each day would examine them all as they ripened. Meanwhile he would find the place on the ground, damp and cool, where he would store a supply first for his mate and then for himself.

Every branch in his trees he would make his own beyond question, by knowing all its turns, its twigs and lichen, and the leaping distance from other boughs. With that experience he would be able to chase out any squirrel that might intrude into his range. He would bound swiftly and safely when he raced with a chickaree friend of his, and when he drove away jays that came into his tree to steal mushrooms.

Those jays! The Chickaree in his nest had been regaining his balance of spirit; the tensions of June eighteenth had been slipping away. But now all his outraged feelings were rising again. All his angry little sounds were pressing to be let out of his throat.

Suddenly he turned his back to the entrance and started kicking. There went the jay feathers out of the hole. There went most of the nest bed, even some of the twigs. He could replace those; send that jay flying!

Sniffing over the floor of the nest, the Chickaree found that every feather was gone. He drew himself into a ball that his tail would cover, pushed his nose into his fur, and relaxed his paws. The tree swung him lightly. Soon he was asleep.

WHAT HAPPENED TO THE BLACK BEAR

As THE SUN shrank the shadows on Beetle Rock, it gave life again to the granite-hoppers. The insects were as still as stone splinters until the heat sprung open their lemon wings. Then they began their angular flitting. A coral king snake loosened his coils so that the sun's warmth would reach more of his flesh. The flesh of the chipmunks and chickarees had its own warmth, but they took the brightness into their spirits.

A male bear pushed out into the steaming grass of a meadow and stayed until his shoulders were saturated with heat. But the sunshine seemed a threat to the Black Bear, a mother of cubs. When it woke her, in the nook where she slept, alarm shaped itself on her face even before she had opened her eyes.

She had chosen this green den because it was well concealed by a thicket. But she and the cubs had been away for five days. During that time the sun had found a break in the leaves. At the unexpected light in the nook, the Bear lifted her head from across her forepaw. With an effort she focussed her eyes. Ferns were shading the brown cub, but a spot of brightness, paw-sized, moved on the black one's side as he breathed.

At the Bear's back was a sugar pine log. From behind it she heard a sound of scattering gravel. Again her ears caught the dry tumble of rocks, and again. She knew its meaning—another bear was digging into a ground squirrel's burrow. Without snapping one fallen twig or leaf, she raised her huge body until her head topped the log. Beyond was a granite slide. The Bear drew her eyelids together as if its bald brilliance hurt, and let her nose tell her what she wanted to know. The other bear was a female, therefore no one to fear. But she reminded the mother that Beetle Rock was a gathering place for bears, including males with their dangerous dislike for cubs. At this loop in her rambling, the Bear must watch her young as alertly as a hunted creature, as a deer or even a mouse.

The mother turned back in the nook. She moved with a troubled vagueness. Except for the log, the walls of this hiding place were but small protection. They were formed of oak leaves, a delicate tangle of mountain lilac, and clusters of pine needles—walls that would hold off a glance, but not a paw nor a muzzle.

The Bear looked to the left, then the right, but not as if trying to see. She lifted a forefoot and put it down again. She shifted half around, doubtfully, unable to grasp any clear impulse. She could not relieve her alarm by thinking. She could not analyze the menace she felt, nor plan new ways to guard the cubs. Instinct and experience had laid an uneasiness on her; she could only wait until they removed it. She let her legs fold beneath her, and lay again in the hollowed loam of her bed. Drawing a hind foot onto the rim, she licked from the sole the dried juice of grasses crushed by it.

At the end of one lick her ears reached for the splash of gravel. It had ceased. Within her thicket, she reared to full height, forepaws on the log. She still smelled the other bear and heard her breathe, but what was she doing? The mother must force her imperfect eyes to see.

A brown blur took clearer shape. The bear outside, tired of digging, sat leaning against a boulder. She was waiting for the squirrel to come out of its hole. Now and then she squinted at the burrow; most of the time she turned a wagging head to the cliff, the canyon, and the sky. She wanted nothing, feared nothing. Human creatures seldom climbed down the steep face of Beetle Rock. If one did, she would hear the shoes on the granite. She dreaded no other animal. And so she rested, with an unchallenged and aimless ease—and with a hint of arrogance that made the mother suspect she knew that she was being watched.

The single bear let herself fall to one side, towards a large mushroom, fragrant beneath a pocket of needles. She clawed it out and fondled it as she ate it, stretched on the ground, dropping her muzzle into her paws for each musty bite.

At the end she licked up the crumbs with a sort of defiant laziness, her brown head weaving interminably above her paws. Was she trying to torment the one who watched?

A temper had begun to flame in the mother's eyes. Her ears lay back and the hairs of her ruff bristled. Now she saw the single female get up and start shambling towards a meadow below. She came to a young pine and rode it down, with the trunk between her legs. She had an air of being too indolent to walk around it, but when the leading shoot was under her mouth, she lay upon the tree and ate its tender top. She turned over on her back, letting the tree spring up. Then she started to the meadow again, her gait loose and her head wobbling.

Suddenly the canyon wind rushed into the leaves above the nook and shook them, tossing a bluebird and a pewee from their perches and swirling a pale mountain swallowtail. As if signaled by the wind, the Bear leapt over the log and galloped after the other female, threatening her with chopping teeth and a rumble in her throat. The other rolled into a trot. But she was not really being driven away. For she slowed down to flip a paw at a seedling fir, as if to express what she felt towards irritable bears. Apparently the mother did not see the insolent gesture. She had stopped and was absorbed in rooting out a lily bulb. An instant later, however, she allowed herself a peek, then an open gratified glance, at the brown rump vanishing among the trees.

This was the other's year of freedom. Next June she would have the cubs and the fears born with them; the Bear would be the one to relax in the sun, to eat or not eat, to follow any brief whim, safe in a world where most animals were anxious. Such carefree security was always the lot of the male bears, but now had come the alternate year when the female, the mother, must be industrious and wary.

The cubs had been born during the late winter, under the snow, in a hollow sequoia log. The Bear's maternal cycle had started. For two months, however, it was a dozing sort of motherhood. None of the family left the den. The cubs were so tiny, scarcely squirrel-sized, that the Bear fed them from stored nourishment in her body. In April, when she brought them out, each was about the length of two paws. By fall they would be half as large as she, and must have learned to take care of themselves. For the mother would sleep off all sense of responsibility before the following spring.

Now in June she was teaching the young ones as strenuously as if she knew to the day how short was the time for imparting all the bear skills. The cubs tried to somersault out of every tedious lesson. First and constantly, the Bear must discipline herself, drive muscles habitually lazy, hold in line a capricious nature. Although safe herself, she never ceased listening, testing the wind, trying to peer into shadows of trees and brush with her half-blind eyes, to protect the cubs from their particular danger.

After the single female left, the mother sauntered around, apparently aimless, actually making a shrewd survey of the neighborhood odors. She wanted to go into the meadow, but first must be sure that the cubs were safe. They still slept. She went a short way up the wooded slope, sniffing into the footprints where Beetle Rock bears climbed to the human village each night. She found no new male scent. At the upper point of the meadow flowed a spring. No male had drunk there since the previous evening. She circled two granite blocks at the base of the slide. They hid no trace of bears, nor did a chinquapin thicket. Finally the mother tested the canyon wind, blowing in over the meadow. Then she walked down in the grass.

This was the smallest meadow she visited—only a pocket of vivid green in the mountain crease between Beetle Rock and the bordering trees. The grasses were thick and throat-high, but the Bear flattened them without effort. A month earlier she would have eaten them; now in June she was moving on to a patch of corn lilies. The naked pads of her feet squelched into mud, for the water of the spring spread over the sedge-roots here; farther down it channelled into a brook, then cascaded over a blade of granite at the meadow's end.

The Bear dug up one of the corn lily bulbs and ate it, crushing it with a single chomp of her teeth. It was crisp and cold, and rich enough to relieve her emptiness slightly. More lily bulbs, also bulbs of blue flowering camass, and shooting stars she scooped out and ate with motions so fast and efficient that they had a gross beauty. Since the Bear was weaning the cubs, her need for food was not the obsession it had been, but she seldom was merely hungry. More of the time she was famished. No taller than a doe, she weighed four times as much and must labor to keep her bulk nourished.

At the ground, the Bear's nose was upon the spice of watercress and the clean decay of sticks and dead roots wet by the brook. The sweetness of Queen Anne's lace and rein-orchis hung over the grass, mixed with the piney warmth of the trees. All of these were only a pleasant blend, the smell of bright summer. More meaningful to the Bear was the scent of mice, and of the Coyote, who must have been hunting them recently. From the thickets below Beetle Rock the wind blew up the scent of a deer herd pasturing there. The smell of deer was often in the air, a dull, familiar torment. The mother caught no trace of the odor she dreaded.

Her weak eyes did not show her the meadow's sparkle—the electric blue of the Steller's Jay, sipping at the spring, the dragonflies' wings, like woven cobwebs, and a western tanager. But the Bear listened to the meadow sounds. One made her lift her head. Was it a click of claws on wood? She waited. Through her misty impression of the meadow spread a quail's call, a sapsucker's drill, and the brook's cool curling over the ledge and crystal trickle below. Then came the voice she expected:

"Muh! Muh!"

It was the brown cub's cry. The tone was a clang, loud and hoarse, as if the voice were a grown bear's, but the expression was helpless, even a little frightened.

The Bear started for the nook. When it cleared in her eyes, she found that the brown cub was up on the log. Still too sleepy to straighten out, he stood with hind legs turned sidewise, forelegs to the front. He was whimpering into the fur of his chest, for he wanted his mother, and food. Instead now he got a push from the nose of his brother. He fell off the log and the black cub climbed up. He walked along on his hind legs, scratching the sides of his belly. But his nose was too high. He stumbled over a snag and fell off, too. The cubs began scuffling. The mother strode up and gave each a cuff that rolled him into the grass.

As the Bear followed them into the meadow, they bounded towards her, their small bodies rocking from front legs to back. They romped against her, trying to force her into a sitting position so they could nurse. She trotted away from them. They thought her escape a game, and rollicked along behind. When they caught her, they tried to nuzzle against her. She spanked them away. Instead of milk this morning, they were going to have a lesson in digging bulbs.

The black cub stood off in the grass, peeking at his mother through the tall blades. She uprooted a lily bulb for the brown one. He turned his face aside. She pushed it nearer. He started to walk away—and suddenly was going much faster than his own legs had carried him. When he stopped, he unwound and whined, not interested in bulbs, very sorry for himself. The Bear dug up two more bulbs and left him looking at them as if they were insults.

Where was the black cub? The Bear sniffed but could not find his scent. A slap of claws on stone drew her to the upper end of the meadow. The cub was on the top of a flat boulder. He had been chasing a lizard, which had darted into a crack. Now he sat up, clinging to a hind foot with each forepaw, teetering back and forth. He would not recognize his mother, for she had disappointed him. Yet—she might, after all, be coming to give him some milk. He tumbled off the rock and ran to her, realized his mistake, and dodged her paw. Then he hid again in the grass.

The brown one too had disappeared. Their hunger would bring them back; meanwhile the Bear would find more bulbs for herself. She started into the meadow, the dusty smell of lichen here blending into the fresh resin of new fir needles, that into a root's dank woodiness, and the pungency of chinquapin. The juiciness of deep grass came into focus, but crossed with a scent less harmless. The Bear stood erect, her nose turned into the wind. Then she called the cubs with an urgent warning, half-disguised, an ominous cough.

They broke out of the grass, too slowly, pouting. The Bear rushed at one and the other, growling through her teeth. If nothing else would frighten them, she would, for they must get into a tree! The cubs bounded to a tall pine.

The mother followed, swimming up the trunk so swiftly that her fur rippled.

The wind continued to polish the grasses. And a purple finch unfurled his song. Twice, and now three times, its last sure note streamed into the sun-bloom. But a huge, blunt head had risen above the meadow's outer ledge. Behind it pushed the shoulders of a cinnamon male bear. He had heard the rattle of claws, and now galloped to the family's tree, agile within his massive bear coat.

With momentum like a wave's against a rock, his immense bulk lifted upon the pine trunk. He stood, gripping the bark and snarling, lips back from teeth long, strong, and white. Between those teeth, any living flesh would be soft.

The mother leaned from the lowest bough. Saliva poured from her mouth as her teeth reached down for the male. His claws tore the bark from the trunk. The snarls of the mother rose to a deadlier clang than his. The male bear embraced the trunk as though starting up the tree.

But his gesture was a bluff. He would have small chance of climbing past the mother, for she would attack from above. His deep animosity was not towards her, but the cubs. Even if he could break through her defense, the cubs would escape him here. She would have taught them to retreat to the ends of boughs too light to hold a grown bear. He dropped with all feet on the ground, but remained beside the pine. A grumble in his throat meant plainly that he controlled the situation, since he could keep the others aloft as long as he stayed in the meadow.

The mother was silent. She crouched closer against the trunk, now seeming to draw within the shadow of her fur.

After a while the male bear wandered towards the grass. The breadth of his shaggy back was impressive, even here near the cliff's great wall. Many animals looked unnaturally small on Beetle Rock, dwarfed by the canyon below them, startling-deep, and the trees above, lifting to unreal heights. Excepting the bears, only the deer and cougars held their size.

But the motions of deer and cougars were as sharply clean as the mountain light, while the gait of the male bear, now, was blurred and undecided. He walked with a pigeon-toed, wavering swing. Often he stopped and waited, blinking.

He dominated only in size, not in action, because sounds and scents, by which he was guided, must be groped for. His attack at close range was strong, but his partial blindness made it impossible for him to chase. Therefore he must be satisfied with food like grass and nuts. The blindness put a curse on his temper, too. He must be suspicious, ready with quick-flaring anger, in case an enemy crept undetected into his fog. No creature would start a fight with him, but some of them tried to steal what food he did get. Sullenness was part of his defense—necessary, but it darkened even his relationship with his own kind.

At the edge of the meadow he waited for something to reach into his boredom. Something did—a thin thread of scent, growing stronger as he followed it to the pine tree the single female had ridden down. All his movements were decisive now, as he smelled around the base of the tree. He partly reared, one forepaw on the trunk, and sniffed up and down the bark minutely. A tuft of the female's fur had pulled out on the needles of a branch. The bear muzzled it as if he could not get enough of her delectable scent. Then suddenly he left the tree and was following her trail out of the meadow. For the present, at least, the mother and cubs were forgotten. The bear disappeared into the grove.

The mother's discipline was not softened by the experience with the cinnamon male. She called the cubs out of the tree and resumed the lessons in bulb digging. The cubs were too hungry now to be playful. They tried the new food. It did not have the lovely milk taste, but felt better than nothing going into their stomachs.

Before their appetites were satisfied, the Bear started the trip of the day. The family would spend the rest of the morning rounding Beetle Rock. In the hollow at the other end, they would sleep until early evening. Then they would descend to the stream and follow it to a greater meadow.

The cubs' training would continue all the way. Black bears ate many things, and the getting of most required practice. Honey, grass, acorns, and berries were found without skill, and the bears stole cones from chickarees. But the cubs must learn to smell beetles under the bark of trees, and roots deep in the soil. Ground squirrels, mice, frogs, and fish each were captured by a different technique.

The Bear had led the way onto the wooded ledge along the face of the cliff. Now came a lesson in ant hunting. The Bear sniffed along the trail until she found an anthill, then stopped with her nose over it. The cubs, following in single file, came up and stood beside her. While they watched, she clawed out a crater. She laid her paw upon it. Soon ants were running over the pads and toes. She licked them off.

The brown cub was the first to try. He put his small paw on the anthill, at once took it off, and began to lick. He got few ants, but he ran his tongue all over the paw, with a pleased hum and frequently a smack.

It was the black one's turn. He sat down on the anthill. The mother cuffed him off and tried to teach him by placing her paw on the sand, then eating the ants she captured. She moved back. Again he sat on the mound. She let him stay. Soon the ants were swarming over his legs and body, even into his ears. He ran around rubbing against tree trunks, and whimpering, but she did nothing to help him.

She walked on. This stone might have grubs beneath it. While the cubs waited, she turned it over. The grubs were there. She licked up one and gave each young bear a chance. A few steps farther the lesson was repeated. The black cub knew what he should do. He saw a boulder twice his size, stood against it, and pushed. It didn't go over! He pushed again, giving his mother a sidewise cheated look. She showed him a smaller rock, but he continued to shove at the big one, now in a little temper. Finally he ran off into the brush. He didn't want grubs anyway, he wanted milk. The brown cub turned over a stone and found nothing but gravel. He lay down and rolled in some crackling oak leaves. The black cub came and punched him. He got to his feet, and the two began to box. The mother left them. Abruptly she trotted off along the trail, keeping her ears turned back, however. Soon the scuffling ceased, and following her she heard the padding feet of the cubs, who had not yet learned the knack of a silent tread.

Now the Bear felt a need to rest and, in her skillful wild way, relaxed without losing any of her caution. Her very motions seemed a kind of sleep, so undriven were they, so lightly directed towards one thing then another. She meandered from side to side on the trail, took the curves roundly, slowed on the upgrades, let her weight carry her down the declines. She stopped to feel the wind, which lifted the sunny-morning heat out of her fur. This shelf on the Rock was narrow, but a strip of wilderness grew upon it.

Trees were near if a bear wanted a refuge, or one could disappear anywhere into a dense confusion of leaves. Here a paw could scarcely be lifted without touching a green spray, or fall without stepping on flowers or the mountain misery, like minute ferns, that covered the ground. Spikes of blue lupine raised their fragrance nose-high. Scarlet hummers' trumpets tossed from cracks in the rocks; the deeper red of snow-plant stalks glowed from the black loam.

The family came to a turn, and beyond it found their path blocked by the single female. She sat in the middle of the trail, tearing apart a sugar pine cone. Since bears take utmost pride in possessing the right of way, an important issue at once arose. Neither female looked at the other, but each would try, now, to prove her dominance.

The unattached bear shifted more squarely across the path; the mother gave all her attention to the cubs, seeming for a change to enjoy their antics. How could she get them on the trail ahead of the other? To circle around frankly would be a humiliation. Suddenly the mother dashed into the brush. This chinquapin bush was an old and hated enemy! Today she would demolish it forever! Standing on her hind feet, she boxed the bush with her forepaws, right, left, and right. The branches swung, the leaves flew. She batted until only a few frayed stalks remained. Then she walked back on the trail, now in advance of the single bear. She called the cubs, with an air of having triumphed over everything, and particularly everyone, in sight. And she really had done that, for a bear who excels in bluff has proved itself the better creature.

Beside the trail stood a bear-tree, a cedar where each who passed must place his mark. The mother stretched up to tear the bark with claws and teeth, as high as she could reach. But an odor on the tree struck the action out of her. It was the odor of the cinnamon male, and it was very fresh.

At the cliff's outer corner the grove came to an end and the ledge broke down into a slide of boulders, polished and bare. Here the animal path, winding among the granite blocks, was marked only by scratches and scents. The mother left it and led the cubs lower, to a den she knew, a cleft in the steep stone mountain.

From behind Beetle Rock, white thunderheads had piled up over the dome of the sky. The wind flowed among the boulders with a sound as if it were on some sad search, and urgent now. The air had cooled; the earth had the smell of a giant fungus. The mother sniffed over the lichen in front of the den. She found no new scents, so the family entered. Already rain had begun to fall.

The cubs stayed together near the opening, standing away from that mother who had good milk in her and would not give it to them. But now she would! She had dropped back into a sitting position and held her forepaws aside. The two bounded forward, climbed onto her legs, and pressed against her body, their ears level with her shoulders. She drew them even closer, with her forepaws on the fur of their hips. As always, the brown cub was on her left, the black one on the right. The black one's claws were deep in the hairs of the Bear's white chest patch.

With the first mouthful of milk the cubs started a small, weird hum, which grew louder as their stomachs filled. It stopped whenever thunder crashed; the cubs would lean tighter against their mother until the terrible sound had rolled away down the canyon, then would begin their nursing again, and their song.

The Bear looked across the top of the black cub's head towards the rain outside, a fine falling that she felt more than she saw. The Rock slanted so steeply here, with no leaves to catch a patter, that the drops struck almost silently near the cave. But their sound rose from the canyon, a great hushed movement like the wary footfalls of all the animals who ever had lived on this mountain.

Suddenly the Bear tumbled the cubs to the sides, got up on her feet, and stood at the entrance. A wind-swirled column of the rain blew on her face. She turned inside again and lay down.

The cubs' meal had filled them with more than milk. The black one, especially, burst with giddy bright feelings. He teetered around on his hind legs, now backing up to try to sit on his mother's sloping side. She reached around and gave him a cuff that tipped him off. Delighted that she would play, he ran forward and began to box with her. Soon she had him down in front of her nose, meeting the blows of his small paws with one of her great ones. The brown cub came into the frolic, and the mother turned over to box with both cubs at once. They squealed with excitement, staggering away, then rushing up to her again. The mother entered easily into their nonsense, which was only the cub form of her own clowning and bluffing with adult bears.

The family played until the mother discovered that sunshine filled the cave opening. Then she ended the game with a show of temper. They must go on, and into a second danger, the vague threat of human beings. Any friskiness would lessen the bears' caution.

The animal trail wound into the hollow between the stream and the human village. The hollow was as close to the cabins as the mother ever allowed the cubs to go.

Perhaps she would not have brought them to this neighborhood at all if Beetle Rock had not been her home during her single years. Then she foraged around the cabins for such ravishing human foods as melons, bacon, and jam. But, tempting as these were, she could resist them when she had cubs to protect.

Human beings, for their part, avoided the hollow. The few who did explore it left their scent, intensified by their fear. The place was one of secrecy and silence. The only green was the ceiling of branches, immensely high. Beneath it the trunks of the trees seemed merely pillars of thickened gloom. Sometimes a snake rustled among dead leaves, but a hawk's cry, faintly sharp in the sky, was usually the only bird voice. Human creatures who wandered as far down as the dry creek bed often were swept with a sudden wish to leave, and would hurry back up the slope, through traps of bone-white sticks and rotting logs. Perhaps the humans sensed the bear odor, without being conscious of it, and were stirred to a primitive terror. Bear scent was as thick in the hollow as the shade was.

For the family's afternoon rest, the mother chose a fir tree midway between the bear dangers and the human menace. She sent the cubs up into the strong concealing branches and lay, herself, at the base of the trunk. Thus guarding them, she dared to sleep.

When the Bear awoke, the light was faint and rosy; the family's day at Beetle Rock soon would end. She took the cubs to the draw along the northern edge of the Rock. From there they would go down the slope to the stream, and on to Round Meadow. As they entered the draw, however, they stopped at a sound of thrashing brush. With the cubs drawn close, the mother waited. The single female galloped by, as fast as if pursued. She vanished into the hollow. Motionless, the Bear listened, and tested the smell of the wind. Then she continued down the draw, now rather hastily. This latest threat had increased her eagerness to leave the Rock.

At the foot of the draw, the family followed a trail worn deep by the feet of bears. It avoided obstacles as water might, turning always to the underside of boulders and brush, around knolls and through gullies with ingenious logic. The Bear and her cubs trotted along without a pause, heads low and swinging. Their hunger had grown with the coolness of the evening.

The animal trail descended to the stream, but a little above the water the bears left it for a human path that would take them to Round Meadow. This was a trail cut into the bank, one strangely level and smooth beneath wild feet.

Now beyond the bears the trail passed a granite buttress, a spur extending out from the slope. For a short way the trail was cut into the rock itself. The mother could see only a few steps ahead at this point but, impatient to reach the meadow, she did not break her speed. Suddenly, on the sheer side of the buttress, she met two men and a boy approaching from the opposite direction.

Bears and humans came to an abrupt stop. They were close to each other on the narrow path. The Bear saw surprise and something like fear on the faces and in the postures of the human beings. But their confidence returned quickly. The leading man said:

"Sorry, bear, but you'll have to get off our trail."

The words meant nothing to the mother, but the tone conveyed what the man had said. It would not be easy to leave the trail at this place. To do it the Bear and cubs would have to leap a distance twice the length of a grown bear.

Why should they, when dominance on a trail was a bear's great source of self-respect? The Bear met the man's look with eyes steadier than his own, but with anger kindling in them. Her ears swung back and her ruff began to rise.

"Come on, hurry up!" said the man.

His tone was commanding. He was trying to force her off the trail. But the Bear did not belong to one of the species, like dogs and horses, that have accepted men's authority. She would intimidate him. She reared, taller now than he, and pulled her lips back from her teeth. She snarled a threat. A degree of real fury possessed her, and swelled into her eyes.

The man's face paled. The sharpening of his scent proved his fear. It was exciting to her! A sense of her great strength swept her, the knowledge that she could knock him from the trail with one stroke of her paw. The man threw back his head. He made a hoarse, defiant sound. He would not turn on the trail. The Bear raised her paw to attack. Only the presence of the second man restrained her. Might he reach the cubs? Even the Bear's pride meant less to her than their safety.

At the edge of the trail was a filling of loose rocks. The man in the rear stooped, picked up a handful, and flung them over the leader's shoulder. Quickly the leader and even the boy began to hit the bears with stones. Human beings had a power which made them superior to bears, even though they were smaller: men could *throw*. They could wound while still distant from an animal. A bear could strike, which few other wild animals could do, but the motion of throwing was not in its muscles.

The Bear leapt from the trail, down the face of the granite spur, into the brush of the stream bank. She called the cubs to follow. The brown cub jumped at once. The black one hesitated.

Now the mother's nose was struck with a new scent. She flung up her head. Striding from the water's edge came the cinnamon male. With a frantic impulse to get the two cubs near her, the Bear roared to the black one. He jumped then, and came down at the feet of the male, who caught him.

The male bear had the cub in his teeth. He was shaking him as a dog shakes a rat. But the mother had reached them. With long claws extended, she was raking into the male bear's shoulders. She was ripping into his flesh. He dropped the cub and turned, facing her on his hind feet. Her jaws were wide as she snarled in a way that should distract his anger from the cub to her.

The male backed away for an instant. His arms were longer than the mother's, and he was stronger, but he was not quite ready to meet a fury so intense. While he paused he was hit on the face with a rock. Astounded, he whirled and discovered two men picking up stones and hurling them from the trail. Another struck him; now one had crashed into his open mouth. He lunged for the white patch on the female's chest, but she was tearing into his side. The male was too confused to fight. He dropped onto his four feet, leapt down the bank, and fled along the stream. The smashing of branches soon was lost in the light sweet syllables of the water.

The black cub lay on his side among the grasses where he had fallen. He was not moving. The mother rushed to him and at once began to look for his wounds. He had a bite on his hip, which she licked gently but thoroughly. Next she cleaned a smaller bite on his side. She took him in her arms and sat on the bank, rocking him and moaning her grief and affection.

The cub opened his eyes. Eager, excited at this sign that he lived, the mother licked his face. The cub stirred in her arms. His consciousness returned quickly.

The Bear called the brown cub and he came out from beneath a mountain lilac where he had hid. Now the mother was anxious to get the young ones into a tree. She took them to a pine, sent the brown cub first up the trunk, then followed with the black one. The tree had strong limbs, two close together where the bears could rest. When the mother had the cubs in this high refuge, she examined her own wounds. She had only scratches on a paw and the tear on her chest, which was under her jaw and impossible to lick; but no matter. Again she put her arms around the black cub. The brown one lay on the branch beside them.

The Bear could hear the human beings starting away along the trail, far below. The male had disappeared downstream. If he returned, the canyon wind would bring the mother his scent. Little by little she relaxed. She would wait here until midnight, when the moon would rise. Then she would take the cubs on to Round Meadow, where they would stop for food, but only briefly. By morning they would be in a deeper forest.

Both cubs seemed to be dozing, when suddenly the black one's paw flew out to give his brother a cuff. The brown one returned a sleepy blow. The mother's paw went down to separate the young bears, but she did not scold them.

WHAT HAPPENED TO THE LIZARD

By June seventeenth the sun was no longer a brightness that merely promised heat. On Beetle Rock that day it had the firm, full strength of summer. It brooded the snakes' eggs, and it dried the wings of the insects, born in a motherless world. It eased the aching leg of the wounded Grouse, and it sweetened the tempers of the mule deer bucks, nervous with their itchy antlers. It prepared the dust baths of the ground squirrels. It made the granite feel as deeply warm as something living.

Perhaps the Lizard received the most direct help from the summer sun. It loosened his muscles and quickened his nerves—as if it said, "Now run, spring up the tree trunks, catch your flies, dart in and out of shadows. Strut the splendor of your throat-spot. Let your bright, inquisitive eyes see as much as many eyes see from one June to another."

The summer on Beetle Rock was still a mountain summer, and it ended on June seventeenth, as every day, at dusk. The cold of winter started flowing down from the snowfields while the sun was still above the western ridge. By the time the sky was black and each star shone with separate brilliance, pools of icy air were lying in the nests and burrows. The birds had fluffed their feathers, and naked young were crawling farther into their mothers' fur. The Rock was lifeless under the feet of the bears.

Even then the blood was warm in the veins of the birds and mammals, in the downy fledglings and the mice with the thinnest pelts. But the Lizard had no way of withdrawing from the cold. His flesh and bones gave up the heat just as the granite did. By dawn of June eighteenth the temperature in the Lizard's crevice was forty-two degrees. And the temperature of the Lizard, too, was forty-two degrees.

Usually the Lizard slept until the sun was high. He woke on June eighteenth, however, while the day was still but a gray mesh over the sky. He had been roused by a jarring of his mossy bed. But apparently no image of snake, owl, skunk, or other enemy had stirred his fear, for he made no shrinking movement and his scales did not grow bright.

The Steller's Jay was screaming that a predator was on the Rock. Presently four other jays flew to its tree to aid with their shrill voices. Birds, squirrels, deer, and many other animals were alarmed by the warning of the jays, but the Lizard's eyes showed only their usual curiosity and trust. In the dim light of the crevice his companion lizard slept.

Softly the light went out. The Lizard turned his head toward the slit where his crevice opened. It was filled with the crouching body of a wildcat. Exposed by the jays as she was creeping to the Grouse's roost, the cat was hiding now until the noisy birds should turn their attention somewhere else.

The Lizard gripped the moss beneath his feet. For instinct told him that death might flash from this mound of gently breathing fur. He watched the cat and saw her paw uncurl. Although he was almost as quiet as the granite, he became more conspicuous with every instant, for excitement made his scales as bright as they would be at the height of his midday warmth.

The cat had eluded the jays. Now they were merely scolding, not attacking, the predator. And finally they were silent. Still the shadow of the huge round body covered the Lizard and his mate.

Then the cat was gone.

The Lizard's feet relaxed, but his eyes remained wide and alert. After such a fright he had little chance of returning to the comfort of his sleep. Soon new sounds proved that other creatures were awake and beginning their day. A chickaree was touring the branches of a Jeffrey pine to see if one cone might have been overlooked. The prick of his feet on the bark was inconceivably brisk. The Grouse was beginning to murmur in her pine tree roost. An explosion of feathers, followed by a long soft whistle, meant that she had gone sailing down the slope. A robin sang with as much contentment as though the sun already beat upon his wings. Only an olive-sided flycatcher seemed to greet the morning with a doubt. His notes came faintly from across the canyon, but apparently they pleaded for reassurance.

The Lizard made no move to join the active ones. He was too stiff even to crawl back farther in the crevice where he could better have escaped the dawn breeze, so perverse in the way it sought out every cranny in the Rock and every hollow in a creature's body. As each lick of its cold tongue sent his temperature lower still, the Lizard closed his eyes. He must wait until the slit of light turned yellow, when he would know that if he forced his legs to carry him outside, the sun would be above the trees, and warm.

When he had felt the worst that the breeze could do, he opened his eyes again. Ahead of him on the moss was the lizard who shared his den. He lay and watched her.

Only a few weeks had passed since she had arrived, unexpected, in his territory. Up over the fallen fir log she had appeared, moving with her tantalizing female jumps, her back arched and her head held low. Instantly a whole new set of the Lizard's instincts had matured. He had thrown himself before his visitor, raising his body on his forelegs to overwhelm her with the splendor of his blue throat-spot. She had not seemed impressed and had started to leave, with an exaggerated bouncy gait. This unconcern only provoked him more. Rapidly bobbing his head, the Lizard had advanced until he was close enough to grasp her neck.

He knew her well now. He knew her female coloring, her dull sides and the light cross-blotching on her back. He knew her passive ways—her willingness to stay within their own plot and her tendency to run instead of fight when another lizard trespassed. She was comfortably familiar to him, or she had been until he watched her on the morning of June eighteenth. On that day she looked different. She looked uneasy. He sensed that she was suffering, for he had an animal's quick awareness of abnormal health.

The Lizard lay and assimilated his misery. Meanwhile the canyon burst with promise of relief for the six-inch creature in the granite crack. Already the sun had lighted the top of the opposite ridge, had lifted out the upper rows of pines from the layered mists. Soon the shadows would fall away.

Sunshine would fill the valley. Then it would stream across the trees on the eastern edge of Beetle Rock. The oval leaves of the manzanita brush would turn from gray to yellow. Splinters of light would slide along the wind-stirred needles of the pines. Specks of hornblende in the granite would flash in the Lizard's eyes—at last he would be warm! Then he would be prepared for any danger and any pleasure that might spring upon him.

The Lizard crept from under the ledge. Stiffly he moved along the fissure in the rock. His appetite would not be keen until he was warm all through, but already he was hungry enough to be interested in a ladybird beetle clinging to a stem of cliff-brake. He could have reached her by jumping only twice his length, but he was not eager for lively action yet. So he noted where she was and let her continue her sleep a little longer.

He had come to a pocket of crumbling needles and shreds of cones and he burrowed partway under them to wait for the sun. He would have liked to doze, but there were so many quick claws clicking on the granite, so many cut twigs falling and woodpeckers tapping, that he was kept alert in spite of his low temperature. It had risen anyway to fifty-eight degrees.

The sun was climbing behind a sugar pine. When it reached the thinner foliage of the top it broke through, and all the terraces of Beetle Rock were washed with cheering color. The Lizard came out from his shallow burrow and lay on the bare granite. Layer by layer his body was penetrated with warmth. His muscles regained their spring. He blinked, and flipped his tail. He lifted himself and drank some dew from the cliff-brake fronds. When he saw that the ladybird had begun to crawl up the stem, he gathered himself for a jump.

At that moment the ladybird dropped to the ground. She drew her legs into her shell and lay still, feigning death. Dead or alive, she was equally appetizing; besides, the Lizard was familiar with that trick. He took her in his mouth and slowly chewed and swallowed her.

When the ladybird was partly digested the Lizard started a trip around his territory. A few steps brought him to a ledge, where he made an exciting discovery. Below, on a sheltered terrace, lay two human beings wrapped in blankets. They were waking from their sleep. The woman was quiet but the man threw off his blanket and sat up.

The Lizard was so delighted to have something new occurring that he felt sharp and quick all over. He ran down towards the strangers, mostly keeping out of sight but watching them from his slanting path as well as he could. He found some ferns in a granite cleft where he could hide and yet have a close view of the humans. They were slow and stiff until the man stood up and stamped one foot that was asleep. The movements of the man were quicker after that, so the Lizard became more cautiously alert. As the man began to fold his blanket he gave it a snap, which startled the Lizard and he flashed back farther beneath the ferns. When nothing else frightened him, he crept partway out again.

Now both the man and woman were on their feet and the Lizard sensed that they were going away. The little creature came out from his cover and streaked across the granite, close to the strangers. He stopped among some dead leaves and looked up at the people, willing to let them come a few steps nearer. Soon he would dart to a new nook, and then to another, keeping always just beyond the humans' reach. But they did not even know of the exciting chase. Still talking, they walked away, down towards the rim of the Rock. If they saw the Lizard, they gave no sign.

A feeling of apathy seemed to sweep over him briefly. He stayed among the leaves until he had lost it. When he came out he was ready to begin the work of his day. That was the double task of filling his stomach and patrolling his territory.

On all Beetle Rock there was no other homesite quite so desirable for a lizard's way of life. The den was in the center with a Jeffrey pine above it. The roots of the tree kept the air in the crevice moist and encouraged the growth of moss. In pockets of soil outside the crevice grew edible green plants: gilia and pentstemon, fennel, and a clump of Sierra asters.

The territory extended in each direction about as far as the pine tree's shadow. This was not a large territory for a lizard, but it was large on Beetle Rock, where rival lizards were willing to fight for each slab of the sun-warmed granite. On the canyon side the Lizard owned a wide, smooth terrace, and towards the forest border, a valuable gully.

Foraging in the gully was both safe and profitable. At the shallow end a lizard could take quick refuge in the litter under a low-branched manzanita. On the sloping sides he could dodge into granite cracks or under the fallen fir log that slanted into the cut. And in the bottom of the gully numerous little crannies opened among the scattered boulders.

Spiders and leafhoppers, beetles, giant carpenter ants, smaller ants, and other insects swarmed over the fir log and through the debris of weathering leaves, pine needles, and pollen cones in the hollow of the gully. They furnished a dependable food supply for the Lizard and his mate.

There were other gullies on Beetle Rock, but none containing a decaying fir trunk. Neighboring lizards still made an occasional bold run down to the log. The one whose territory adjoined on the east seemed unable to give up the impression that he might yet win the gully. In one battle over the question he had lost his flexible tail. That had happened in April when the territory boundaries were being decided. By June eighteenth his rigid substitute tail had grown more than half an inch, but the challenger's good sense had not increased accordingly. Once a week at least, he came back to be taught with a few new nips and bites that the owner had no intention of moving out or of sharing the gully with another male.

The Lizard started the survey of his territory, as usual, at the lower border. He would be glad to add to the ladybird when he had a chance, but he was not ravenously hungry. He passed a golden-mantled ground squirrel sitting on a rock-pile, a pair of chipmunks chasing each other, and an Audubon's warbler seeking insects on the edges of an incense cedar. These animals shared his territory, as he did theirs, and all accepted each other. What he would not have endured was the presence of another lizard, but he found none and gradually came in towards his gully.

He had a moment of terror when he saw the shadow of a bird slipping over the Rock and realized that he was far from cover if the shadow should be that of a red-tail. He froze, hoping to avoid being noticed, then sensed that the shadow was moving faster than a hovering hawk's would. The shadow of the hawk was usually somewhere upon the mountain, but must now be darkening other granite fields, or wing-splashed trees, or the grassy shelters of the mice.

Carpenter ants were parading over the log in the Lizard's gully, but they were common and easy to catch and he was in a mood to hunt. Soon he had found a spider spinning itself down from a branch of the manzanita. A leap and a snap and the Lizard had it in his mouth.

The Lizard liked to be quiet and warm as long as he could feel any food digesting, so he stayed beneath the manzanita for a while. The air was still within the tangle of branches, and the sun came through almost unbroken by the vertical leaves. From the dead brown needles, twigs, and leaf-mat under the Lizard's nose, the heat drew a dusty, spicy odor.

When he felt the spider becoming a part of himself, the Lizard left the manzanita, for a ladybird and a spider did not make a morning meal. He licked up several of the ants and worked along a crack in the fir log, where he found a larva of a sow bug and ate that. At the lower edge of the log, in a pocket of loose bark, he saw the Deer Mouse hiding. She was a stranger in his territory. The two pairs of eyes met, appraised and accepted each other. The Lizard jumped down into the bottom of the gully and lay among the leaves. Several easy victims crawled in front of him and continued to live because they were not appetizing.

The Lizard resumed his hunt. First he examined a growth of fennel on the side of the gully and next some clumps of grass along the rim. He ran from one green plant to another, coming ever a little closer to his eastern neighbor's territory. Finally he dashed across the border, into his neighbor's patch of golden-throated gilia. He found neither food nor excitement there, however, so he returned to his proper grounds.

Now he began hunting from one end of his territory to the other, covering the boulder surfaces in leaps. The sun heated him from above and the granite from below until his muscles were so limber that it seemed as if he could outrun any other creature. Released from his trap of cold, he flashed

about with the brilliant movement so stimulating to his kind—as though hunger for speed as much as hunger for food were the impetus for his search. The terraces, like the gully, had cost him many a fight, but now that he could claim them, his greatest pleasure was to race upon them.

What finally tempted the Lizard's taste was a leafhopper, a gamey, delicately tart green hopper. She was on a scarlet pentstemon, nibbling the edge of a leaf. All the Lizard could see from below were the tips of her working jaws. He leapt to take both leaf and hopper in his mouth. At the touch of his foot on the plant, however, the hopper jumped to a ledge above, and the Lizard missed her. From where he landed he could not see where she had gone. He sprang onto the ledge himself, but the hopper had disappeared.

The Lizard stood and looked around him. He had been too eager, had leapt too carelessly. Now his bright spirit seemed all collapsed. A small ant twinkled towards him on its red legs and the Lizard licked it up. But the ant wasn't even a giant carpenter.

A scream struck suddenly through the air. The Lizard darted to the nearest crack in the Rock. There he lay, wary and still, as he listened to the Weasel and a golden-mantled squirrel battling.

They were on the slab in front of the Lizard's hiding place. Twice he saw the whirling, clutching bodies cross the granite. The squirrel was familiar to the Lizard, for it lived in a rock-pile in his territory. At first its shrieks and the snarls of the Weasel told that the fight was going evenly. But the squirrel's cries grew more desperate, while the Weasel's ceased. Then there were no more cries, no sounds. The warm stone and the sun were like the stone and sun of a dead world.

The silence flowed past. The Lizard closed his eyes, for the sharpness of his fear had left him tired. Finally the song of a robin rose in the thicket. Gradually it loosened the voices of other creatures.

The Lizard crept out on the open granite and relaxed, at right angles to the sun. The heat seemed to reach his very heart. Yet there was a limit beyond which it would not be safe to let his temperature rise. For a while he kept it down by lightening his color so that his scales reflected, rather than absorbed, the sunshine.

As he became lighter he became brilliant in color. Green stripes appeared on his back and around his tail. The yellow on his sides was bold. If he had raised himself to impress another lizard with his throat- and belly-spots, they would have flashed an iridescent blue.

In beauty as well as energy, the Lizard was the sun's creature. He must wait for its touch before he could fully live, and must give up his liveliness when the sun withdrew. But now its heat was becoming dangerously intense; even the Lizard's color-control could not protect him from it. He had begun to move towards a shady crack when the granite ahead seemed to lift in a gray wave, which came sweeping forward, catching him in a swirl of sand and dust. The early breeze had died but a new damp wind was rising.

Above the treetops had appeared a higher mountain, hugely soft and white. It spread so fast up the sky that it could not be missed by even a small, prostrate creature accustomed to watching the surfaces level with his eye. Soon after the Lizard noticed the cloud, it hid his sun.

Now he must find a nook quickly, for raindrops soon would be slapping the Rock. A lizard caught in any downpour might be swept away, over the Rock's rim, into a brook that would carry him to the river. At least a sheet of chilly water was sure to come streaming down the granite.

The Lizard hastened towards his den, running on the tips of his slender toes, making a single line of flowing silken movement, his tail held out so lightly that it gave to every leap a finish of perfect grace.

He came to the grasses near his crevice, circled them, and when he had reached the opening looked inside. His mate was gone. He entered. Though his companion was missing, something was there, something new—a pair of eggs, each nearly as large as his own head. They were lying on the moss.

The Lizard never had seen eggs before. Cautiously he went over to them and touched them with his tongue. They were soft. He darted away from them, across the den, where he lay in a far crack holding off the strange white objects with his eyes.

When the eggs continued to remain quiet, the Lizard dared to turn his attention to the crevice opening. Being a lizard, he regarded his mate as a permanent companion. She no longer excited him, but she pleased his eye when she moved and added to his sense of comfort when the sun went down and they crept to the back of the den to sleep. The gathering of the storm would normally have caused her to return, so that her absence now was as strange as her discomfort earlier had been.

She appeared in the entrance as the first drops fell. The Lizard watched her while she found herself a soft bed on the moss inside and relaxed, curling her tail up her side with a look of delicious ease. She ignored the eggs.

The crevice was dry and snug, for it sloped enough to prevent the moisture from draining in. Sometimes the rain splashed as it hit the granite near the opening. The Lizard blinked when it struck his face; otherwise he kept his eyes on the pelting water, which came down faster and faster, breaking into a fine spray on the rock. The fennel outside kept dipping under the heavy drops, then springing back. And lightning flashed, brightening the den. The Lizard's scales grew light again because the lively movement was so exciting to him.

The storm slackened. For a while the Lizard could still hear the din of rain and thunder, then even that disturbance ceased. The sun came out and the only sounds were the water dripping on the granite and the small horn of a red-breasted nuthatch, piping its one note from the forest wall.

The Lizard left his crevice. Vivid and clean were the chickarees' piles of red cone scales, the yellow and violet plates of the pine bark, and the grasses, and green leaves of the manzanita, chinquapin, and mountain cherry. The Rock steamed in the patches where it was not yet dry.

The air was becoming warmer, but the Lizard waited for a while beneath the fennel. Its stalks bent over him, enclosing him in a little sky of bright wet stars.

When the sunshine was sure again, the Lizard ran up over the Rock to the highest point in his territory. This granite knob was his favorite lookout for hopper, snake, hawk, or lizard-challenger. These were slow in appearing after the storm, but the Lizard's temperature, and with it his energy, were rising as the air became warm. He lifted himself on all four legs and compressed his sides so that his throat and belly bulged and made him look tremendously impressive. When no male lizard happened by to be intimidated by this fine display, he relaxed on the surface of the granite.

Yet he was restless. Other animals were coming out of their retreats and in their various ways were expressing relief at the storm's passing. They were delayed for a short time by a commotion near the Weasel's den, but as soon as that was over a robin chased a chickaree through a fir, one of the deer marched up the Rock with an exceedingly high step, and the jays dove through the pines in slanting, motionless flights like blue sunbeams.

The Lizard moved a little way along the ledge. Next he ran back and sprang on the trunk of an incense cedar. He circled it a few times with increasing speed, dropped off, and went bounding over his widest granite terrace. Now he was flying about with an almost angry swiftness, darting in one direction, then, when he'd caught his breath, whirling back, as if he were dispatching an intruder for the final time. Gradually he worked his way towards the eastern border of his territory.

As the Lizard and his neighbor had settled the question, their boundary lay along the edge of the pollen cones fallen from a ponderosa pine. The trunk of the pine was in the neighbor's grounds; the Lizard's came to an end at the outermost of the cones.

This day, however, the Lizard proceeded straight through the cones to the base of the tree. The Grouse was picking up the pollen capsules, breaking out their yellow dust wherever she stepped. The Lizard glanced at her, then paid no further attention to her. He leapt on the trunk of the tree, ran up a short way, then down and back to the ground. Suddenly he saw that his neighbor lizard was watching from the bed of gilia. As if his presence on the tree had not been impudent enough, he challenged the neighbor by throwing out his throat and belly to display his blue spots.

The neighbor dashed into the circle of cones. Instantly the Lizard was back on the tree trunk, spiralling up, down, and up again, a whirling streak, always out of reach, always by quick genius on the opposite side from the owner of the pine. Three times the chase took the pair as high as the branches, the Lizard always leading. On the next descent he jumped to the earth. His neighbor dropped beside him. Immediately the two were skirmishing among the pollen capsules. A yellow cloud hung over the ground, so that neither could see distinctly, but still they found each other and escaped each other.

Dry leaves from an oak tree had settled among the cones. They too were flying. Their crackle was attracting attention all over the end of Beetle Rock. Among the eyes that looked to see what was happening were a coral king snake's. When the snake found two lizards absorbed in fighting each other, he slid rapidly forward.

The Jay was watching the battle, but he did not fail to discover the snake. He screamed a warning. The combatants did not listen, and knew nothing of their danger until the snake's head struck.

The head came down between them. Both lizards tried to dart away. More than once when the snake had captured a lizard, his teeth had gripped only a discarded tail, so now he gave all his effort to catching the one with the stump, the owner of the tree. When this victim found himself in darkness he tried to work free, but the snake's curved teeth were hooked into his scales, and the jointed jaws moved forward steadily to enclose him.

The Lizard was very content not to win the fight with his neighbor. As soon as he saw the snake, he raced away to the safety of his gully. There he stopped in the leaves beneath the manzanita to regain his breath. Later he moved up cautiously to his den. His mate was lying at the entrance and he stretched out near her on the granite.

For the rest of the afternoon the Lizard was quiet. His instinct now was to get himself in balance—to recover his energy and compose his nerves. He let the sunny day go. His short months of wakefulness were at their peak, but he gave up the granite sparkles, the spiders, the lively chipmunks, the wavery white butterflies, and other delightful, keen sensations, and allowed himself to be carried away from them in sleep.

As the light was becoming mellow, hunger roused him and he returned to his gully. He did not try to catch any jumping insects but he ate a few of the ants.

When he came back to his crevice the sun was slanting into the opening. He crawled inside and lay on the moss, out of the breeze, yet in rosy warmth.

His companion came in, too. After the sunshine left the den the temperature dropped quickly. Degree for degree the lizards became as cold.

Soon they were asleep. The end of their day was a chilling and numbing like the gradual loss of life—at the time when swallows were skimming the richest insect harvest from the air, when chipmunks raced as boldly as if the sunset were a cover, and the ears of the Mule Deer flicked to catch every twilight-sharpened sound.

WHAT HAPPENED TO THE COYOTE

THE NIGHT OF JUNE SEVENTEENTH was a time of hollow luck for predators. The wildness of the wind had made the mice, hares, flying squirrels, and other prey unusually cautious. The same wind put a fever into the hunting of coyotes, weasels, cougars, owls, even the shrews. They were searching recklessly and catching little.

The Coyote on Beetle Rock paused towards midnight to recover strength and to wait for the moon, due to rise from behind the eastern peaks. As he stood upon an angular boulder, thrust into the canyon from the cliff, the wind parted his fur. The wind was pouring over Glacier Pass and Mineral King, builder of thunderheads. It was flowing down and past the Coyote like a movement of the night itself. Beyond the mountains it had swept across the southeast deserts and Mexico's high plains, home of the small wolf breed. The smell of the mesquite flowers still was in it.

The Coyote shivered. His tail was tight along his right flank, not blown there, but pressed forward as was the Coyote's way, part of the impulse that made him stand with feet close, throat pulled into his ruff, and brows lowered to the pupils of his eyes. He was a creature that would draw his body around him in utmost concealment. But he was too sensitive to escape thus from the impact of scents, sights, and sounds. His trembling was not alone from cold.

Now the moon was up. Small clouds were blowing across it. Their shadows upon the earth were dark shapes flying over the wooded canyon slopes. The oak leaves in the thicket below the Coyote sharply cut the air, like a myriad of bats' wings. The Coyote turned. He would climb to the Rock's rim, there throw back his head and cry out his own wildness, send into the wind and the moonlight streaming between the clouds the breath of his inaccessible spirit.

He turned, but as he started away he heard through a moss-grown crevice of the Rock a muffled yelping. Below, in a hollow formed by the tilted corner of a granite block, was the den of the Coyote's mate and pups. One had wakened. He was arousing the others. The Coyote listened, a forefoot raised, ears down towards the crevice. The sounds he heard were not good. In the yapping was a whimper which told the Coyote what he knew already—that the pups were close to starving.

Instead of climbing the Rock, the Coyote trotted towards the meadow where he had foraged earlier in the night. During several hours of hunting there, he had caught but one newborn field mouse. The other mice of the litter had escaped, clinging to their mother's teats as she fled through the grass. The Coyote had cached the tiny prey while he continued searching, but finally had taken the one pup's mouthful back to the den. Perhaps now in the moon's light fewer of the mice would escape him.

The Coyote's den was near the west edge of Beetle Rock; the meadow was at the other end. He rounded the cliff on a wooded ledge, along a bears' trail, warily glancing from side to side without a break in his light-footed lope. In the angle between the cliff and the forest of the mountainside, on a small bench, lay the meadow. To enter it the Coyote must cross a sugar pine log. He could have jumped it, but for caution's sake climbed over, gripping into the bark as deftly with hind feet as with forefeet, swinging down his ears when his head came level with the top. On the other side he circled right so that the wind would carry his scent into the canyon.

The meadow was bright enough to promise better hunting. The moon, shining white on the granite wall, reflected a sheen down over the grasses, broken where the bears had waded through. This was the bears' meadow. None was here now, but their scent almost covered the scent of the mice. The Coyote's nose moved quivering in the air, as if tracing a web. He could distinguish the odors of meadow mice and deer mice. Only luck would give him a deer mouse, for the routes by which they sped from nook to nook were unpredictable. But the meadow mice had little roads trimmed through the grass roots, leading direct to the home nest, or to an enemy's mouth.

The Coyote walked, one slow step at a time, out into the grasses. He brushed down into the stems with his nose. Soon he found a mouse's runway. He sniffed along it, putting each foot down gradually before he rested his weight upon it. Now he had reached the end of the runway, where the small path entered the ground. He waited, tail slightly lifted, nose pointed, eyes sharp on the hole where he hoped soon to see a softer shadow.

Few coyotes bred in the mountains would have tried to stalk in grass so tall and thick. Instead they usually pursued their prey over open ground, or dug it out. The creature now desperately hunting food for his young had spent most of his life in the foothills. Since he was inexperienced here, he sought the mice where their scent was strongest, in the meadow.

He had been in the mountains only three months. He and his mate had been born on a ranch beyond the western spur of the canyon. Besides the wild prey, easy to catch there, they could kill a sheep whenever their hunger became acute. They had raised one litter of pups successfully on the ranch; the next year two of their young had been poisoned. The following year they lost the entire litter by gunfire, traps, and strychnine placed in carcasses. The Coyote pair, then, had migrated to the east, where they found no signs that men attacked wild animals.

They traveled over the crusted snow on an old road to the crest of the Ash Peaks Ridge. Beyond they saw a country unfamiliar to foothill creatures, of deeper canyons and loftier mountains. They went on. When the road joined a highway cleared of snow, the Coyotes struck away from the human trails, straight up the steep, forested mountain. They climbed higher and higher, trying to find more level ground.

They reached it only at the top of Beetle Rock's perpendicular cliff. The tumbled taluses of the cliff itself provided many a hiding place for hunted animals. In a morning's search the Coyotes had found a den for their spring brood.

That decision made, they had set about catching unknown kinds of prey in this world of trees and rocks. Few wild animals ever survived in a country so unlike their birthplace. Some other coyotes had done it, but even by June eighteenth this pair had not yet proved that they could. The Coyote's hunting had a hint of panic in it, for he was driven by the hunger of all, the pups, his mate, and himself. Since March they had been approaching starvation, and for three days now there had been nothing in his stomach but a little grass. The cavern of his ribs seemed to hold only a craving. Pain stabbed in his joints. His eye sockets burned.

The grass in which he waited for the mouse was taller than he. It brushed the sides of his face and bent under his belly. In the foothills, sheep had cropped the grass close; the Coyote could speed over it easily, and anywhere see to the roots. His method of stalking was adapted to that shorter grass. Here he was fairly drowned in the limber fibres, and his stalking motions, decisive and fine, so effective on the lowland range, repeatedly failed to capture the prey. But now a freshening of the mouse scent speeded the Coyote's heartbeat. There was the lighter cast of fur in the mouse's burrow entrance. Backward swayed the Coyote's body, down a little went his haunches, ready for a spring. At a movement in the hole he was in the air, forefeet descending stiff, with all his weight behind them. But even in the instant of his pounce the grass had swung together and warned the mouse. No chunky bite of food was under the paws that struck. The paws flew right and left among the grass stems.

The Coyote leapt forward, then his angry jaws raked back along the runway. Once more—how many times that night!—he had lost the mouse.

Down the center of the meadow flowed a brook. After stalking and missing several other mice, the Coyote caught one at the edge of the water. At once he perceived an advantage there and waded up the icy stream, his nose in its borders of wet roots. Three more mice had been added to his catch by the time the birds awoke and began their tinkling chorus of chirps. The Coyote carried all but one of the mice back to the den and left them beside a stone where his mate was accustomed to look for his offerings. More tired even than hungry, the Coyote then went to his own den at the base of a snag, and slept.

Although he had eaten far too little food during the spring, similar amounts probably would have kept him alive a long time if the food had been captured for him. But now starvation had slowed his speed. Unless he learned more tricks of mountain foraging soon, he might be unable to catch any prey at all. Hunting required that one's powers be keen, even in country that was familiar.

When the Coyote awoke he started a tour of the Rock's borders. Many squirrels and chipmunks had dug their burrows there where the forest broke away into thickets, brush, and the gray-leaved, bright-flowered gravel plants. The Coyote seldom had caught a chipmunk, whose agility the swiftest predator must respect, but occasionally his pounce had captured a digger squirrel. The Coyote crept along the lee side of the brush, scenting, watching, listening.

He was moving down the draw north of Beetle Rock. Across the dry streambed he saw a squirrel-sized hole. The hole was plainly visible to any animal passing along the draw.

The mound of earth beside it drew attention to it, an animal home very different from a coyote's burrow, which always had an entrance hidden by brush or rocks, and the excavated earth so scattered that even another coyote could not detect it. The hunter crouched on his belly behind a weathered log. Through a gap in the wood, broken by bears' feet, his gaze searched into the hole.

A ground squirrel's gray furred head appeared. The Coyote pulled his legs farther under him, ready for the spring. But the ball of the squirrel's nose started an alarmed quivering. The breeze had swirled and brought him the Coyote's scent. Now the squirrel was sending forth a piercing whistle, a warning repeated at precise intervals, loud enough to reach every animal near the Rock.

The Coyote's chin went down onto the gravel and he closed his eyes. But quickly he lifted his head again. His wits had pricked through his discouragement with a plan. Warily he backed along the log, then circled down through the shaded hollow behind the draw. He was going for his mate's help.

At all seasons except when the mother was guarding young, the Coyotes hunted together. They had worked out strategies that captured many creatures neither could have caught alone—jackrabbits, whose single bounds were six times a coyote's length, and deer, and other prey. One of the Coyote's difficulties in getting established in the mountains had been the mother's absorption in the care of the pups. Were they not old enough now to be left alone while she assisted with the hunting?

The Coyote trotted out onto the boulder above the den. On the granite shelf in front of the den were his mate and the six young coyotes. Normally they would have been tumbling and scuffling in the sun, developing in play the

sharpness of mind and body that they would need when grown. The mother would have been wrestling with them, giving them affectionate nips to teach a coyote's flashing attack and guarded defense. But the father looked down and found his mate and four of the young ones stretched on their sides or curled with noses in their tails, motionless. One pup was licking a sore on a foot. The sixth stood at the shelf's edge looking into the sky that filled the canyon, taking his hunger with the lean stamina of an adult wolf.

The Coyote made his way down a cleft in the boulder, his sides shrinking narrow below his hipbones as his shoulders dropped at each descending step. The pup at the edge bounded to him hopefully; two others raised their heads. The Coyote went to his mate and pawed her tail away from her face. Sensing that he had brought no food, she growled annoyance at being wakened. The Coyote took the long fur of her shoulders in his mouth and pulled. Not understanding his plan, she let herself be dragged across the shelf, still showing no intention of getting up. The Coyote barked. Then she knew that he wanted her to follow him.

The two loped off the end of the shelf, up through a thicket and the wooded hollow to the draw. The squirrel, lying half in its burrow, piped its shrill warning still. The Coyote's intention was as clear in his partner's mind now as in his own. The pair, keeping out of sight, came into the draw above the squirrel. Now the Coyote's mate bounded towards the squirrel, openly and apparently with confidence that she could catch him. When he shrank back into the hole, she continued down the draw slowly, letting her tail hang limp, her ears droop, and her shoulders sag. She went on to the den while the Coyote waited, concealed behind the log. The squirrel did exactly what the Coyotes had expected—peered out, saw a discouraged wolf leaving the draw, felt safe, and came out of its hole.

The watching Coyote was so excited that his breath fluttered. He slid forward to make his bound through the break in the log more sure. Now! He tensed his muscles, but suddenly a dizziness blurred his eyes. He shook his head cautiously and his vision cleared. The squirrel was beginning to dust in the gravel. Again the Coyote gathered his strength. He hurled himself over the log. But he had misjudged the distance. He landed a paw's length short of the squirrel. It dodged—and vanished, perhaps down another entrance to its burrow.

At the den his mate would not be sleeping now. She would be waiting, watching feverish-eyed to see the Coyote come loping in with the fat squirrel, meat for all of them. He did not return to the family. With a blind homing impulse, a tormenting urge to be again in the brushlands, he loped down the surface of the Rock, past its upper and lower fringes of manzanita, into the oak thickets below, through dark groves of cedars and open stands of pines and firs rising obliquely from the steep slope. On down he went, down in the direction, at least, of the ranch. There would be prey that he could catch even though he was weak. There were the short grass and the low shrubs, the smooth, wide range with its long visibility. Lower and lower trotted the hungry Coyote.

He came again to cliffs of sheer granite, polished by ancient glaciers. In the clefts a few trees and grasses had taken root-holds. The Coyote wound down from ledge to ledge. But when a pocket of fallen needles on which he stepped slid off the smooth rock, and he nearly plunged into the canyon, he took his way along horizontally to a more gradual, wooded strip of the mountainside.

He followed a deer trail used by the herd on their yearly migration. Few other tracks marked the slope; the dense animal populations were lower and higher. Many logs and dead branches, hurtled down the mountain during floods, lay, white, among the tree trunks. They all pointed downward and the Coyote continued in their direction. But in crossing one log he paused on top. He had sensed a change in the air.

Clouds were massing over Mineral King. Colors in the trees and earth were changing, becoming darkly vivid. Many sounds had ceased; others were unusually clear and meaningful—the chirping water of a spring, wind approaching in the trees, cones falling through branches and striking the ground heavily, with an answering deep, muffled ring from the mountain. The air smelled of damp dust and the charred wood of snags burned in a forest fire many years before.

The Coyote's mind began to link up memories of other storms, images of the rain's end, of animals coming out of their holes at the return of the sun. That would be a time of good hunting, probably in the very draw where he had lost the squirrel. The Coyote dropped from the log and started up the slope.

He had climbed halfway back to Beetle Rock by the time the first drops fell. Gradually the rain soaked through to the skin of his muzzle, legs, and haunches, and he began to shiver. But he continued up the mountain. When he came to the base of the Rock he turned west, and continued to the gorge at the Rock's corner. The stream, already swelling from the rain, roared out of the gorge and down the side of the canyon with a sound of great vitality. The Coyote followed the stream back up into its dim ravine. At one place a sequoia had fallen across the water, forming a bridge. The Coyote crept under it where the bank was dry.

Around him was a thicket more dense than any he had seen in the brushlands. If he had been well he might have felt too covered in it, but now there was comfort in the sheltering of the great tree trunks, and the leaves of hazel brush, ferns, lupines, and grasses, darkly glistening with moisture. The wet evergreen needles had an odor clean and pungent to a feverish nose. The stream's white foam looked cool.

The rain falling on the leaves, and the churning of the water, were the only sounds. Bird songs and the squirrels' calls all were silenced...unless that was a bird's peep. The Coyote sat up and peered among the sedges below him. Soon he had discovered a fledgling, a bird he did not know. It was a mountain dipper, wren-shaped, gray, matching the water. On the edge of a stone, with its feet in the current, it continually squatted, down and up, to imitate the ripples and thereby escape notice. Yet it kept reminding its parents in a shrill voice that it wanted food. Except for its piping and the opening and closing of its orange mouth, the Coyote probably would not have seen it.

Wings broke out of the foam. The dipper's mother, who had been foraging on the bottom of the stream, flew up, put a caddis fly larva from her beak into the young one's, and dived again. The fledgling resumed its piping and squatting. The catch looked easy. A single leap brought the Coyote to the dipper's rock, yet he did not get the bird. It escaped him by plunging into the water. The Coyote never had known a land bird who could do that. Once more the mountain world had tricked him.

When he went back beneath the sequoia he climbed farther into the angle between the log and the bank. He pushed in as deep as he could, curling himself with his tail across his right foot and his nose hidden. Usually when he lay thus he kept his eyes as high as the guard hairs of his tail, so that he could see about him. Now his entire face went into the deeper fur.

No creature would find him here under the great log, probably not even a bird or mouse to make him try once more to force new strength into his muscles. A nook like this, remote yet airy, was the best possible place for a sick coyote to hide, perhaps to stay for a very long time.

The Coyote slept—only briefly, but when he awoke nothing remained of the rain but a bright mist. The clouds above the trees were a fresher, softer gray; the air felt as if some wariness had been abandoned. This was hunters' weather. The earth itself seemed lazy, pleased, warm, overconfident. Some new arrangement of natural forces, of tensions and pressures, would make the pursuing kinds of nerves and muscles more effective than the fleeing kinds.

Lying now with his chin on his crossed forepaws, the Coyote looked out with awakening interest. All the leaves were lifting little by little, as their burden of water dripped away. Even the gray wreckage caught at a turn in the bank seemed alive, for the rain had given the branches a fullness as of nourishing sap. The stream had risen enough to form a water-wheel, spray flying off its rim.

The Coyote slid out from beneath the log, yawned, and stretched with his forelegs on the ground, his haunches at full height. Then he trotted towards the animal trail that slanted up the bank.

The predators' luck had changed, but luck never was all of success. As he climbed towards the draw, the Coyote rounded a patch of brush and surprised the Mule Deer Buck and his younger deer satellite. Instantly singling out the younger buck, less experienced at defense, the Coyote flung himself into a chase. Faster, even, was the buck. Away it bounded, the Coyote pursuing, snapping, aiming for the buck's hind leg. Suddenly the deer whirled, faced the Coyote, and struck at him with its hoofs. The Coyote dodged. He was winded. He moved off a short distance and waited, panting, while the buck trotted out of sight. He was not discouraged by this defeat. Even in the best condition, a lone coyote seldom caught a healthy, grown deer.

Finally reaching the draw, the Coyote prowled among the rocks, the last of his energy blazing up as his need for food became unendurable. He found no basking squirrels, but under an oak tree discovered weasel trails, all leading to a burrow. Frantically he began to dig. The wet earth crumbled beneath his paws. The scent was so fresh that he expected every scoop to disclose the nest.

A cry shot up from the gully nearby. Up the side sprang the mother Weasel and several kits. The mother leapt for the Coyote's face, her attack doubtless meaning that more young weasels were in the burrow. The Coyote and Weasel, two predators, met fang against fang.

The Weasel hung by her teeth from the Coyote's lip. She clawed for a hold in the fur of his throat. He flung his head to shake her off, but still she held. With a stronger swing he got free of her, ripping away the flesh she had gripped.

He turned in a slanting lunge, shoulders low, his throat drawn back into his ruff. His attack was in his long pointed mouth, with its teeth ready to strike from lifted lips. The Weasel hurled herself at him again. His jaws snapped, but she evaded them. Finally she darted out of his reach into a hollow log. He dug again into the burrow and soon came upon two weasel kits. With a quick shake he had broken each of them.

At last he knew again the feel and taste of food in his mouth. Although he was starving, he ate in a coyote's lingering way. It was a light meal, yet it strengthened him for a further hunt.

The Coyote loped into the brush below. Now a doe flew past him. She was a yearling, fleeing so wildly, so plainly in a panic, that she seemed likely to stumble and perhaps break a leg. The Coyote raced after her, throwing the whole of his strength into this chase. His coyote's speed, fastest of all wolves', might have equalled the doe's on an open hillside, but here he fell behind. For the deer was keeping a straight course, bounding over the top of brush, logs, boulders, and seedlings, while the Coyote must circle most of them. Soon he had lost even the hoofbeats.

But he did not give up that doe. His memory turned again to hunting in which his mate had assisted. He lay on a rock until his panting quieted and strength came again into his trembling legs. Then he went back to the den.

His mate seemed more alert; perhaps she too had found food. She trotted forward as if she knew his intention and together they started out. He led. As they climbed over logs or threaded brush, drew up to tracks, sniffed them and galloped away, they moved with a harmony which hinted that coyotes had a talent for a paired life. Only wolves and foxes, among all wild mammals, mated permanently. One reason why they did was suggested by the shared intuition that made the Coyotes hunt well together.

They were following the fleeing doe by scent. She had crossed the stream, then climbed the opposite slope. On that side was the same kind of country, a forest broken with brush and boulders above the same long cliff of which Beetle Rock was the knobby end. The trail took the Coyotes to a meadow. Was the deer somewhere in it, lying hidden by the jungle-like grasses? Keeping well back among the trees, the Coyotes started around the edge. They found the doe on the other side.

She was eating a mushroom which had broken up through the fallen needles under a fir. The Coyotes would have marked her anywhere as prey, for her jerky motions showed nervousness apart from any sensible fear. She constantly shied, prancing with arched forefeet; a grasshopper's click was enough to make her leap aside. Between bites she threw up her head, with whirling ears. The Coyotes were adjusting all their movements to the wind, for if the doe caught a whiff of their scent she would be away and choosing her own direction.

From now on the Coyotes intended to control that doe. Soon they would start her off, the male Coyote at her heels, his mate a short way ahead to turn her up the slope or down. If she acted as the deer did in the foothills, she would flee in circles. That would give the Coyotes a chance to chase her in relays and keep in fresh strength themselves. While the Coyote was running the doe through the first half of her circle, his mate would trot across and be ready at the halfway point to take her turn. He would trot back, leisurely, and relieve his mate at the starting point.

Now the Coyote was considering whether the upper or lower slope was more clear. This was their first deer relay in the mountains; later they might learn instead to run the deer into the rocks, down stream banks, or over cliffs. The chase they planned today, a foothill strategy, would not be easy, but both Coyotes were eager to begin it; leg and chest muscles, ears, even tails were tense. The Coyote decided on the upward course, turned to signal his mate—and saw a cougar, poised for a spring at the doe. It was directly above her, among the upturned roots of a fallen sequoia. Within an instant the great cat leapt, striking the doe's back with all its weight, bringing her to the ground lifeless.

Already both Coyotes comprehended that the cougar's appearance need be no misfortune—that it even might solve their existence in the mountains. The cougar could provide them with a livelihood, at least until they were more at home here. One often had done that in the brushlands. They had not known that they might meet the same involuntary benefactor in their new environment.

As was its custom, the cougar dragged the carcass away from the scene of the slaying, downhill, this time to the grassy bank of a brook. With extreme caution the Coyotes followed, then watched the cougar from a knoll above it. To their noses came the scent of fresh meat. They saw the cougar enjoying such a meal as they had not tasted for several months.

The sun dropped below the branches of surrounding trees. Now slanting between the trunks, its light fell on the cougar. A pair of ravens began wheeling above. At sight of them the Coyote shifted his forelegs with anxiety, for other coyotes were likely to see the ravens and come loping in for a chance at the carrion. If he must, then, he would fight for this meat. But no other coyotes appeared and finally the cougar was satisfied. It buried the rest of the carcass, as the Coyotes expected, scraping leaves and needles over it. Then it walked away slowly across a rocky clearing, the tip of its long tail upturned and slightly swaying. Beyond the clearing it vanished, shoulders first, apparently over the edge of the cliff.

The Coyotes allowed themselves only the briefest meal here. Many times a cougar did not return to a carcass, yet it might. So the Coyotes began at once galloping back towards the stream with pieces as large as they could carry swiftly. They hid the meat in a log. After they abandoned the cougar's cache, they cleaned their muzzles and chests on the grass, then took their stock from the log to the stream, across and up to the den. Each pup had a great chunk of nourishment, and at last the parents could stop and fill their own empty stomachs.

Most of the meat would go to the pups and their mother, for the Coyote must leave to follow the cougar, to trail it on its circular migration. There was no question about the success of its hunting. Regularly there would be cached carcasses, which the Coyote would transfer to new hiding places. The Coyote family would be assured of a food supply until they might hope to perfect their own strategy of mountain hunting.

With no ceremony of leave-taking, the Coyote started out from the den, to pick up the cougar's trail and shadow the cat for the six to ten days before it would come back to Beetle Rock. It had not returned to the cache. The Coyote found that it had wound down the cliff, crossed the stream, circled the base of Beetle Rock, and proceeded along the canyon side. Apparently it was heading for the greater adjoining canyon at the foot of the Western Divide.

Most of the sun's light was gone, even from the sky. Bats were cutting in and out among the tree trunks, and the forest seemed suddenly full of robins, all on the ground, hopping through the dusk with a sociable, soft liveliness. None bothered to flutter away when the Coyote passed, for they sensed that he was not foraging and he was no enemy at other times. He did not kill for pleasure.

The cougar's trail came to the intersection of the canyons. Here the Coyote paused, for the trail was becoming almost too fresh. He climbed to the saddle of Moro Rock, from which he could see across both canyons—on the left to the ridge of the Western Divide. A faint flush still tinted the snow of the crags, but it was fading, and they were becoming even more remote. The foot of the range already was lost in darkness. The rosy peaks seemed to belong more to the sky than to the earth.

The Coyote threw back his head, closed his eyes, and began his song. It sounded as though it came from several throats, for it was blended of different strains, beginning one after another somewhat as brooks formed by the snow of the crags joined to become a stream and a river.

Apparently he was not calling to anyone, or asserting anything but his coyote nature. The song was as lonely as the crags, and as wild as the water's cry.

WHAT HAPPENED TO THE DEER MOUSE

THE DEER MOUSE was trapped by a sound. Startled as she sped about on the floor of the night, she had run beneath the edge of a stone. There now she hid in her fur, nose on her chest, feet all covered. But her ears' wide membranes stood high with alarm, twitching as though attached to the sound by threads.

She heard a whisper of something coming. Something was sweeping on thin wings between the needles of the trees; something was streaming across the hillside grasses, brushing the brittle husks of the seeds. Feet were stealing, seeking through the dry stalks, and leaping over the Mouse's stone.

Sometimes the sound sighed down. Then the Mouse would dare to turn around, trying to fit deeper into the crack between the stone and the ground. She wanted the walls of the nook to press her all over, but however she crouched, one of her sides had no touch of shelter on it. That side yearned with a sense of lack, with a sort of skin-hunger, quite apart from its feeling of cold.

At other times the sound snapped. Then the Deer Mouse dropped flat, ready to dodge a hunter's pounce. She did not know that the enemy was less substantial, even, than she—that it was the wind, tossing twigs and rustling the grasses. Most of her experience was in matters as small as seeds and flies, the voices of mice, and the look of her clean white fore-paw with its tiny claws, like pearl slivers. What could she understand of a battle of winds above a canyon?

The Deer Mouse wished to return to her mother's nest, an earthy cavern among the roots of a manzanita bush. The nest-ball of grass and feathers was as soft as the arching belly of a mother mouse, and behind it the roots were strong, like a mother's bones that would hold off any enemy. The Deer Mouse had grown beyond the need for cuddling, but she longed for the proof on her skin of sheltering walls. Not tonight alone, but for several days, she had been obsessed with the loveliness of crannies shaped to cover a mouse.

Shortly before midnight she could have run back to the nest quite safely, for there were signs that the real dangers had drawn away. The owls were down in the hollow, as their hooting told, but in tones too low to sound in ears of deer mice. The bears had gone up the draw and were not yet due to return. Now the coral king snake would be too cold to forage. The swirling wind carried no warning of wildcat or shrew. But the swish in the grass had more meaning for a mouse.

Finally the sound shifted farther off. The Deer Mouse left the stone and raced to the manzanita, each bound as light as if she were a wooly aster released from the wind's pressure. She entered the nest and crept upon the furry heap that was her mother and sister. Now her back and sides were pushed against the dome of fibres, giving her such solid, real assurance of being hidden that she quickly relaxed in sleepy peace.

Her vibrissae were brushed by a fourth mouse groping in. He was the brother, who had left the family several days before. The wind had brought him home. Its uneasy sounds had meant for him, too, that predators were swarming through the forest, and made him long for the security he associated with his mother's nest. But while he was fitting himself among the other mice, the earth jarred. A powerful weight had struck beside the bush. It was a great horned owl who had seen the brother slip into the dead leaves over the roots.

The leaves were swept away and claws scraped into the soil. Then the nest jolted as the roots were pulled apart. The whole pile of mice was a-tremble with frantic little heartbeats. All the mice were ready to leap as soon as the nest was torn—that was now!

The Deer Mouse alighted beneath a pine tree and darted into the fallen needles. Around her was a fanning out of quick patters. That shriek must have meant that the owl caught one mouse, perhaps the sister, always slower, more cautious, abnormally afraid of predators. The Deer Mouse's nose was pushed into the soil, but her great black eyes looked up through the needles. Against the brightness of moon-silvered clouds she saw the owl rise and sweep away.

The needles were no firm shelter. Now the Mouse was bounding down to a log in the crease of a gully. She crept under the log's curve and waited until her panting ceased. Then she found a split in the wood where she could hide. It was a tight refuge, but after such a fright she could endure being crowded by the touch of walls. The middle of the night had come, a heavy time when deer mice liked to sleep. She let her fears fall away and closed her eyes.

Moonlight woke her. It had entered even her cramped little niche. The Mouse slipped out and down the dark side of the log. She was hungry, but the log was surrounded with glittering granite gravel. The very sight of it made her back up tighter against the wood. Not even a tuft of grass grew near, for the log lay in a cradle of sand. But its shadow extended beyond the sand. The Mouse prowled out and found a spot where the soil smelled tempting. Her forepaws whirled into it, scooping it towards her so fast that she was continually astride a mound, which her hind feet kicked back. Now her claws struck an acorn, split by its swelling germ. She hooked her sharp front teeth into it, gave it a mighty pull, and it came free. She ran back to the log, then, with the acorn in her mouth. There she ate it, morsel by morsel—delicious nut, succulent with new life. The Chickaree had buried it. His stores, cached around the gully, had furnished the deer mice with many a meal. They watched even his cones, and when one dried enough to open, they reached between the scales with their tiny paws, drew out the seeds, and carried them to their own hoard in a pine stump.

Feeling untidy now, the Mouse licked her forepaws and scrubbed the pink tip of her nose. Taking each hind foot in her forepaws, she turned it and washed it. With the hind feet she smoothed her shoulders, and she washed her sides and back. Even her striped furry tail she cleaned. A quick shake then, and all her pretty coat, white below, oak-leaf brown above, was smooth. She seemed hardly at all like a house mouse, more like a toy doe, and so similar in color that it was clear why human beings had named her a deer mouse.

She had slept, she had eaten, and she was groomed. With these needs out of the way, she became a quiet mouse, facing a great emergency, for she was homeless. As she crouched in the shadow of the log, some delicacy of pose, or soft wildness in her eyes, gave her the unreal look of fawns. She was a small, brief union of breath, pulse, and grace, yet the apparent nothingness of her, the hint that she soon would vanish, if she had not already, actually was her strength.

All mice were so hunted that some kinds had become erratic, but deer mice learned to bound ahead of the strain of attack. They lived with a lightness that served very well for poise. Always—their way was—be ready to drop the game or gnawing or nest building, and disappear, noiseless as a blown thistle. Dig to the sprout with airy speed, before a falling leaf may warn of claws on the bough above, before the breeze flies ahead of an enemy bringing his scent. However close come the beak or teeth, then, never hoard the fright. Don't let even death be important, since it is so familiar. Try to escape if there is a chance, but if there is not, give up life quickly.

Yet the Deer Mouse was more than a fluff of a little being. She, as well as any bear or coyote, must have her established place at Beetle Rock. Among the boulders, brush, and trees must be one cranny recognized as hers. The wrecking of her mother's nest had made it necessary to find her own niche and her own life, but she was ready, anyway, to cease the play of a young mouse and become a grown one. Before her brother left, he had raced with her on the boughs of the manzanita, and she had loved that swift motion as fawns love to bound down a slope, or chickarees to leap from tree to tree. Afterwards she sometimes had run through the bush alone, whirling herself exquisitely half out of her senses, but now a different interest had stirred in her, an impulse that soft speed would not quiet.

Oh, where, now, was a cranny into which she might fit—some hole sweet and snug, with firm walls, a secret entrance, so placed that winds would not blow, nor moisture drain, into it? Lowland deer mice dug burrows, and in some other places they made nests in trees, but the Beetle Rock deer mice searched for their nooks instead of building them. The Mouse would begin at the log.

She found a knothole and started into it, eager and pleased until she discovered a scent of other mice. She turned down towards the ground. A strip of bark, loosened by beetles, hung away from the wood. The Mouse crept under and felt the space, its size and shape, not with her paws or nose but with her vibrissae, the whiskers of various lengths which she moved like fingers over an object to give her information about it. Around this pocket went the spray of her tactile hairs, quivering into every crack. At the far end she turned. The wood and bark touched her all over, as she liked, yet some instinct warned her that this could not be a permanent home.

Now she looked up the moon-spangled gully. Almost irresistible was her wish to return to the thicket above. The shadow of a cloud fell upon the gravel, and gave her the chance to go.

Back in the manzanita thicket, the Mouse slipped under the dead leaves beneath. As she pattered along, ears flattened and vibrissae down her sides, she was hidden from any searching eye, although the leaves kept whispering:

"She is *here*."

She was making her way towards one of the root-crowns, but when she reached it she found that a family of juncos lived at the center of that bush. The Mouse crept close to their cup of woven grasses, so softly that the mother bird continued to sleep. She climbed up on the rim of the nest. Most of the junco's feathers were the color of a night shadow, but the sides of her tail gleamed white, and so did her ivory beak. The Mouse touched her tactile hairs over the wings and tail, outspread to cover four nestlings. She was sniffing, too, scenting the birds' light breaths, and their flesh, delicate with the sweetness of seed-food. Once more she felt over the strong walls of the birds' home, over the mother, and the nestlings' down at the edges of her wings. Then the Mouse dropped again into the brittle, rustling leaves.

At an opening in the brush she came to a mound of the leaves, pawed together by a deer. She scrambled through them, but they all flew; nothing here was strong enough to support a nest. The next manzanita bush had been home. The Mouse sped past. A short way beyond, she came to a braid of scents left by the Weasel and her five kits when the pack of little hunters crossed the thicket. The Mouse dodged away towards the open ground under the trees.

Her home-range extended on one side as far as a spring, and on the other to the foot of an open slope. In the space between grew three pines and an ancient fir. The Mouse knew the exposed part of the trees only as circular trunks and a vague overhead thicket; trees, to her, were underground things. She had her own view, too, of the earth's surface. That was not the smooth mat of needles it looked to a human eye. Chains of small shelters led almost everywhere.

Here was this tumbleweed, only a tuft of dry twigs, but a mouse could hide under it. One leap from there and she reached a fallen branch. The tiny foliage of staghorn lichen covered it. Beneath the lichen she ran to the other end, her feet spinning like the feet of a house mouse. A bound, then, to a piece of bark, and from that to a root of the fir—she must let this search for her nook take her wherever it would. Most animals looked in particular kinds of places for their home-sites, but a deer mouse's cranny was accidental, an earth pocket washed out by the rain, a stump just enough decayed, a log fallen aslant a rock.

The cloud passed; moonlight slid out upon the forest floor. Now the Mouse must go underground, down into other animals' burrows, hoping to find an abandoned one. Deer mice often did appropriate such homes, whose owners had disappeared, having suddenly, unexpectedly, no more use for the patiently dug tunnel, the nest chamber, and the storeroom filled with seeds.

She would begin with the burrows of the meadow mice, who lived in the grass near the spring. She crept into the stems, moving over a web of runners and dry fallen stalks. In it was the ground-litter of this tiny wilderness—petals of the grass flowers, seed husks, skeletons of dead insects, and living insects, sleeping or numb with cold. When she found these, the Deer Mouse stopped to nibble them up, while the wavy currents of the grass stirred above her, the sound in the moist green blades as harmless as the song of gnats.

Before she had gone far she reached one of the meadow mice's surface roads, which led to all parts of the grass patch, in a curving network, regular as a cobweb but more graceful. Underfoot was a smooth pavement, kept clear by the industrious owners. The Deer Mouse passed two at work on the roads, chewing off new shoots which might be food in some places but here were troublesome weeds. She went into each side-path, and soon found one that ended at a burrow entrance.

Down she sped into the clean little tube, just mouse-width, now straight, now curving to pass a root, a path all dark but impossible to lose. With the smell of the soil was mixed the meadow-mouse odor, grass juice crushed into fur, and musk, and here an extra odor, that of milk. The Deer Mouse followed a branch tunnel to a nest chamber. There she found five newborn young. They cried to her with appealing squeaks; perhaps they thought her their mother. She felt all over the little mice, then returned to the main burrow.

She saw mice eating, carrying nest fibres, and sleeping, for meadow mice worked either night or day. At the opening of each tunnel she sniffed the mouse scent and everywhere found it fresh; there was no vacant home for her in the grass patch. She was not anxious, anyway, to live where a road through the stems might lead snakes or weasels direct to one's nest.

Each time she had left a burrow, she had shaken the dirt out of her fur. Now she washed all over. When she felt clean she was ready for a new exploration.

She bounded to a break in the matted fir needles, but it led to no cave, only to the emerging stalk of a snow plant. She crouched there briefly, above the roots of the tree. She knew that a crook of its largest root was sheltering a chipmunk family, parents and young, whose burrow was the cleanest of all the underground neighborhood. It had even a separate space for empty seed husks. Twice while the owners slept, the Mouse had prowled through their home, one to envy but never likely to be hers. Chance hardly would remove seven chipmunks at one time.

Meshing into the roots of the fir, too, was an underground village of digger squirrels, a labyrinth of hiding places, nests for families and single squirrels, and places for the stores of seeds packed in dry sand. Sometimes the squirrels deserted old nests, but the Deer Mouse would not investigate now. Morning was too near. The waking squirrels would not be friendly to mice.

Among these tunnels were others that the Mouse did not know. Her last search took her into one. The entrance was a well-concealed hole between stones. Beyond, the Mouse crept down and down, much lower into the earth than she had been before. Finally the tunnel turned and wound beneath the tree's root-platform.

Strange and remote was the smell here, of very old soil, powdered fibres of ancient plants, and the dust of rocks. Even the roots above the tunnel smelled of the past, for the tree's food and drink came now from newer roots, pushed out into fresher humus. The Mouse knew the odor of the occupant; she had smelled it behind the lively heels of the golden-mantled squirrels. The air in the burrow was very cold, but the Mouse, delicate though she was, could hold away the chill. In fact, when the winter snows would fall, her little white feet would be running over the frosty white crystals, while the owner of this home, much larger than she, would have retreated into sleep.

The Mouse reached the nest chamber, where the squirrel lay sleeping, coiled in his fur. From here several tunnels radiated to a passageway that half circled the nest—a whole web of roads for escape. The Mouse went into a tunnel beyond and abruptly found herself in the open air. She had come out at the foot of the fir trunk, through an emergency exit disguised in the bark.

Now daylight was lifting the night away. The ground had a yellow-gray cast on it, too bright for safety. The Mouse would spend the day behind the loose bark on the gully log, not a trustworthy niche, but the best she knew. She started towards it. But why did this strange deer mouse lie on the needles, warm, alive, yet surely not sleeping? The Mouse found the hole in its skin, smelling of the poison transferred with a shrew's bite. And here came the shrew, smaller than the Mouse but with venom for her, too, in its pointed snout. The Mouse escaped the death-sleep; practice in racing had made her feet faster than the shrew's. Soon she had curled up between the bark and the log, her panting already becoming lighter, and her round eyes narrower, now but gleaming black lines, and now lost in fur.

A few times she roused, but only briefly, until late morning. Then a strange sound reached in and loosened her sleep. The wind had shifted and was driving the sand in gusts, like minute sharp rain, against the bark. Soon real rain was dropping in coarse thuds on the log. All damp things smelled strongly of their dampness.

The Mouse pushed back against the walls of the niche. The rain was becoming a roar. A sudden thunderclap startled her to a blank. Her ears had not begun to rise again, nor her forepaws to relax, when a stream of cold water began to flow over her tail.

She climbed higher in the nook, and the water rose. She crept along to the end of the bark. A turbulent brook now entirely circled the log. The Mouse did what she must—plunged into it and spun over the top. Soon the current whirled her against a branch. She raced back upon it to the gully-side, then bounded in pelting rain up through the thicket, to the trunk of the fir and the burrow exit of the golden-mantled squirrel. In his tunnel she shook herself vigorously, made herself as flat as a turtle, and pushed into the soft soil of the wall. Here her own heat would make her dry. She fell asleep.

After the Deer Mouse woke, she slipped out into the tunnel and, perched on the sloping floor, washed and groomed her fur. No place more practical than this cranny could be found for a deer mouse's nest. But of course she could not stay; a golden-mantled squirrel would not allow his exit to be revealed by the path to a nest of mice. The squirrel already had smelled her there. She heard him coming up to drive her away, so she slipped out into the daylight.

Now sunshine fell in most places where the rain had fallen, making the earth steamy. The Deer Mouse crouched between the fir trunk and a fallen cone. Her eyes were flicking over the ground, trying to find a better shelter, when the Coyote came prowling from under the thicket. He walked towards the fir.

His scent was almost as tangible as the pierce of sharp teeth. How keen was his hunting; he was not living lightly; for him this instant might have been the storm's peak, so intense were his nerves. His nose was at the ground and his feet moved forward compactly. He was following a scent trail, perhaps of the Mouse herself.

Of course he would find her here. She had no chance to escape—yet she must make that last, desperate leap. But when the instant came that she would have jumped, the bit of buff shadow lay instead a trifle lower between the cone and the fir trunk. The Mouse had slipped into a faint, perhaps thus saving her life, for if she had sprung from her refuge the Coyote certainly would have seen her.

When her speck of consciousness drifted back, the Coyote was gone and the Mule Deer Buck had come under the tree. He stood looking across the gully, apparently also aware of the Coyote, for his feet shifted in a strained way and his head was high, his ears pointing stiffly forward.

What a great, powerful creature he was—yet he shared with the Deer Mouse a fear of their common enemy.

With the Deer there, the Mouse felt more safe. He would not harm her. When he relaxed, she ran out from her hiding place and examined branches and tufts of lichen torn down by the storm. One might be large enough to give her a refuge. But while she searched, another deer came, a doe in so nervous a temper that soon the Buck's hoofs and the doe's were stomping wildly. The Mouse must leave.

As the Grouse fluttered down from one of the fir boughs, the Mouse looked up. At once she raced for the trunk. In a hole up in the tree she had seen the face of a flying squirrel, a gentle creature that she had met at night on the ground. Now the Mouse reached the hole and stopped, clinging to the bark. The squirrel turned its mellow eyes upon her. The Mouse crept over the rim of the hole and down into the nest cavity. At the bottom lay four young flying squirrels, piled together as the Mouse had slept with her family. She pushed between the fur of one and the bed of shredded cedar bark.

During the afternoon the screams of a red-tailed hawk woke the innocent creatures in the tree. There was a brief, startled stir as each tried to creep in deeper, and then they lay still. When the fearful cries came no more, the squirrels moved slightly, easing their tension, and the Deer Mouse slipped farther into her furry refuge.

The next time she opened her eyes she backed out of the squirrel's coat. Now she was sharply awake. The hole at the top of the nest cavity shone but faintly. She ran up. The night's darkness had drawn to the western horizon, but the star-brightened sky cast a soft light into the trees bordering Beetle Rock.

On a branch outside the hole the flying squirrel

crouched, ready to glide to the ground. Looking like a furry leaf with her legs and their connecting membranes spread, the squirrel dropped into an air current that took her lightly to the ground. The Deer Mouse ran down the trunk of the tree. Families of deer mice sometimes shared the nests of flying squirrels, but the Mouse still preferred to find her own niche. Besides, she had no family.

Perhaps tonight she would explore the slope above her home-range. The air was so still that no stalk or leaf stirred against another. She nibbled through the bases of several lupines so they would fall, and then ate the succulent tips. When her hunger was satisfied, she made herself dainty and neat, and bounded off into the grasses.

Near the top of the slope the grasses thinned, finally coming to an end. The Mouse continued on among the trees. Ahead she saw the speeding, white, upcurved tail of another deer mouse. He led towards something that was new to her—a cabin, a nook of human beings. He ran into its open door and the Deer Mouse went in too.

The human creatures were not there. But upon everything lay their scent, the scent of predators but not of animals that preyed upon mice. The Deer Mouse did not belong to the unclean species of mice who lived in the dark corners of human homes, but she felt no terror here, only her natural wariness in a strange place. The other deer mouse had climbed at once to a shelf and was gnawing a box, apparently familiar with the cabin and not frightened. The Deer Mouse began to explore.

Everywhere she found corners, and they gave her confidence. Few were completely covered, but they were nooks that she could back into and feel the shelter-touch. The room itself had corners and there were others around the shelves, luggage, books, and many more objects, mysterious to a mouse. She liked the fact that nothing moved. Since hunters must always move to catch prey, mice's eyes were alarmed by most motions, even of leaves. Here all was pleasantly still.

The Mouse smelled and touched many curious things. While she was examining a cold metal flashlight, she seemed suddenly overwhelmed by all the strangeness and sat up, clenching a small forepaw against her breast and quivering her ears to find the other mouse. His gnawing had ceased. The Deer Mouse raced down the table leg and towards the door, but discovered her companion eating at a pile of oats which had poured from his hole in the box and onto the floor. She stopped and tasted the oats. Delicious! Now she sat beside the other mouse, rapidly nibbling pawfuls of the new food.

Outside, sounds approached—a man and woman talking, walking towards the cabin. They entered and shut the door. The Deer Mouse zigzagged across the floor, hunting one of the room's crannies. Dimly she saw her companion's tail slipping behind a dustpan. She glided along the edge of the wall and joined him.

Now the darkness was destroyed, not gradually as when the dawn comes, but instantaneously. The man exclaimed over the spilled oats, came striding towards the mice, and lifted the dustpan away. Silent as shadows the mice moved behind an ax. The man swept up the oats and threw them into the stove, hitting the dustpan on the iron with a noise so sharp that both mice winced.

The people's voices were tremendous, and might have been frightening, but curiously were not. For all their loudness, they had no angry tones, as animals' growls did. The Mouse could hear only the higher tones, and therefore caught more of the woman's voice than the man's. Both people were getting undressed, and the mice watched. After the human beings had taken off their shoes and part of their clothes they sat on the beds, across the room from each other, still talking.

As abruptly as the light had come on, it went out. There was a creaking of springs, shrill in the Mouse's ears, when the people got into their beds. Their voices continued. The other mouse returned to the food shelves as soon as the room was dark, and the Deer Mouse followed. Now he was gnawing a new hole in the box of oats, for the man had turned the other hole to the top. The Deer Mouse helped with the gnawing. The man said:

"Do you hear something?"

Then both people were silent, and the mice, too, kept quiet. When the voices began again the mice chewed once more at the box. Soon the oats were spilling onto the shelf and the mice were eating them.

Finally the people talked no longer. First one, then the other, breathed more deeply and slowly. For the male mouse this was a sign that he could make more noise. He began to gnaw at a crack under the door. He was a little knot of energy, now flat, his mouth turned straight up as he chewed at the bottom of the door, now huddled against the crack so that nothing of him showed but his furry haunches, and now a mound that pivoted from side to side while his hind feet kept a steady grip on the floor. While he worked, he held his tail straight out, its tip upturned with eagerness. But occasionally he became tired or bored, and pattered around the cabin.

A large splinter came away and the male mouse slid out. The Deer Mouse followed. She leapt from the sill to the ground airily and ran to the base of a tree, where she crouched. The other mouse came out of a burrow hole under a stone, and bounded towards her.

Lightly she sped away from him, back down the slope and into the grasses, dodging among the stems in a bewildering way. She seemed to be trying to lose the fine small rustle that followed, yet she never ran so fast that she quite escaped. Soon she was in her home-range, leading over familiar ways.

This was like racing with her brother, only somehow more amusing. The Deer Mouse dipped under a root, up and over another root, a gasp of a little run. A pause, then as soon as the pursuing patter came close, she rippled across the gravel to the log. She slipped to the log's other end. Finally she led to one of the manzanita bushes.

Back and forth on the gnarled stalks the Mouse flew, as if she had snapped the threads of gravity. Even the lift of a bird's wing hadn't a freer motion. The other mouse raced well, too. Sometimes he would leap from behind her onto the branch above or below, and skim ahead. Then, unexpectedly, he left her, ran down the main stalk and crouched at the base of the bush, a secret place only visible from the branches above. The Deer Mouse could see him there. And he could see her, the gleaming white fur under all of her body, streaming along the boughs.

The male mouse beat on the bush with his forepaw, a spray of patters, a pause, then another quick knocking. The Deer Mouse crouched motionless and listened. The drumming was repeated. She came in along the branch. Once more he drummed. She hesitated—then sped down. With a soft brightness she drew up to the mouse. Playfully she began to nibble his ankle.

WHAT HAPPENED TO THE STELLER'S JAY

IN THE TREES' FLUFFS of shadow that night were many small, bright, black points, the eyes of birds wakened by the conflict of winds. The wind reached the coast as a single north-flowing stream, but the Sierra ridges divided it. Above Beetle Rock two of the currents were meeting again. One had swept down from Mineral King, unopposed until midnight; then the canyon wind strengthened, rising against the slope. The colliding gusts battered the trees.

Birds like the flickers and chickadees, in the hollows of rigid dead snags, were not roused, but the wind alarmed every bird on an outside perch. It was giving a wild ride to the Steller's Jay, clutching the end of a limber branch. And no bird could fly to a firmer roost until morning, since wings cannot grope their way through the dark as feet can.

The nest of the Jay's family had been abandoned weeks earlier. Now on June eighteenth, the Jay's young, almost grown, perched among twigs near their father. They were better able than he was to keep their hold on the jerking branch, for the Jay was tattered in plumage and energy from the long strain of rearing the five. Even yet, though they were stronger than he, he hunted most of their food.

If a whole crowd of jays had been competing in jauntiness, the bough's movement would have been a chance to show off balancing skill. If the sun had been lighting the Jay's performance, and one observer had watched, his spirit might have been stronger. But no jay is fitted by temperament to struggle alone in the night's obscurity. Finally the exhausted father was almost ready to let the wind snatch him, toss him above the trees in a last, light, effortless flight, then fling him down on the granite. Still his toes gripped the perch. At the most sudden pitchings he fluttered his wings to stay upright, but he never quite lost his touch on the bough, which was his touch on life.

At last the dawn separated the trees. As soon as it clearly showed new landing places, the Jay flew to a lower branch, which extended into the sheltering boughs of a pine.

The pine was the Grouse's roosting tree. The Jay could see her, a motionless knot of shadow close to the trunk. As he sat and watched her, his wings and tail flicked. He was impatient to have her fly down for her day on the ground. She stood ahead of him in the forest boss-order, and had dominated him so many times that he knew he was not supreme in the tree with her there.

But now he forgot her. Suddenly he was taller, alert and hopeful. He had heard a flake of bark hit the earth, and a scratching of claws. Oh, let that be a predator for him to berate! For many weeks the jays had not molested any animal hunter, because with young in the nests the parents must keep themselves inconspicuous. But the Jay's caution dropped away when the bark dropped from the pine trunk. He dived for the lowest branch.

A wildcat was climbing the tree. The sight of her was like a poison that exhilarates. The Jay screeched taunts at the cat, so vigorously that four more jays flew in from other trees. Each seemed to try to make his cries the most shrill. As the cat climbed, the birds kept pace. The Jay took the lead. He flew towards the cat's face, apparently meaning to jab his long black beak into her eyes. Of course he always stayed far enough back to be out of her reach, but his raids were brave-looking, impressive gestures. Anger flashed from the cat, from her very fur. Once it drove her to dart a paw at the Jay. He dodged away with a clattering clamor.

The cat was nearly as high as the Grouse's roost. Though the Jay was not fond of that bird, he would not let her become a victim. The cat turned to pass a branch that intervened between her and the Grouse. But at that point she paused. It was useless to go on; the jays' din made it impossible to surprise the Grouse, as she had intended. She began to back down the tree.

On the ground she raced away to a manzanita thicket and disappeared. The jays had followed her down but soon were up in the branches again, scanning the Rock, still crying threats after the enemy who had escaped. Gradually their excitement subsided. One by one they became silent, and all but the jay father left the tree.

This was the first time his brood of young had seen a jay flock mobbing a hunter. They had watched the performance without helping to make it noisy, but they had absorbed the spirit of it. When the Jay wished to perch quietly and enjoy his triumph, they flew at him, thrusting their squawking beaks into his face. They were screaming only for food, but now with increased vigor. As the Jay looked at the big birds, whose stomachs never were full, the exuberance which had stiffened him crumpled. Without it he felt very tired. He knew that those cries would continue all day, however strenuously he worked to feed the five.

This brood was his first. He had no experience from which to foresee the steps in his cycle of fatherhood. Guided only by instinct, he had helped his mate to build a nest of new pine needles, softened in water for weaving. He had fed her while she incubated the eggs, had guarded the nest, and when gaping mouths had replaced the eggs, had filled them tirelessly. His mate had helped, but on the day before this she had disappeared. The Jay did not know the reason; he had not seen a goshawk tossing jay feathers down from a tree.

During those weeks in June, the forest clanged with the cries of young jays, dozens of them, old enough to find their own food but demanding that their parents supply it. Curiously the mature jays, so sure in movement, so energetic, could be imposed upon—by many small animals besides their own offspring. Supported by a flock of their own kind, they were bold, but individually they weakened when other creatures became decisive or vehement.

Soon after the wildcat left, the Grouse sailed out of the pine. The Jay continued to perch on an upper branch with his young screeching around him. Suddenly he screamed back at them, surprising them, and even himself. Then he drooped on the bough.

The battle of the winds had been won. Now the canyon current swept up the mountain, steady and swift, as if to meet the sun at the crest. This up-flow, the deepening of the sky's blue, the temperature, and the feel of the branch were a combination that stirred a memory. On a morning in a previous June the Jay, then a fledgling himself, had felt cramped in his parents' nest. He had climbed over the rim to the bough at the side. Never had he flown, but the wind pressed against his breast. It pried beneath the edges of his wings. When he lifted the wings a little, it grasped their wrists and flung them up.

Anticipation had trilled in the Jay as the air flowed under those upraised wings. The invisible support was almost as strong as the branch he grasped. It seemed to urge him to launch out and rest upon it. Then he was in it. Sky was beneath him, all around him. As the mountain dropped away, panic had struck him, but that set his wings in motion, as the sense of drowning prompts a creature to swim.

In a quick surge of learning, the Jay's whole mechanism of flight had begun to work. Unexpectedly his wings were beating the air...backward and up...forward and down. On the swift rise they almost relaxed, letting the wind fling them high, with quills opened like slats. At the top of the stroke came a splendid flip of the end-feathers, then the Jay felt his breast muscles grasp the wings and pull them into the slower downbeat, quills now locked in a solid sheet. Even then his wings had moved gracefully, only three beats to a second, like the great buzzards', while the sparrows flapped thirteen times in the same interval, and the hummingbirds two hundred times.

Until that day the Jay's ribs always had expanded to draw in his breath; when he flew they clamped into a rigid support, and his wings' motion sucked the air through him, through a chain of air sacs under his skin, in his muscles, in even his bones, keeping him cool without perspiration, circulating fresh oxygen in amounts that replaced his strength as he spent it.

On this June eighteenth the similar wind under the Jay's breast awakened a similar anticipation. Again his wings rose involuntarily, and he found himself flying above the canyon. His leaving was so abrupt that the young did not follow. He was alone. Unreachable! Instinct had shown him how to father and rear a brood; now instinct said that the undertaking was finished. By leaving his five, he learned that he might leave them. He was independent, with no more need for discretion, again free to be the flashing blue focus of many eyes, again free to mind every other animal's business.

First he needed food, large quantities all for himself. Too smart to seek it where his young might discover him, he circled the face of the cliff to its east edge and alighted on a grassy bench. Leisure seemed to expand all around him as he stood on the mossy side of a spring and sipped the cold water. The sunlight cupped in the little meadow had warmed the bordering trees. The Jay would forage among their branches, for the sun was almost like nourishment to flesh as hungry as his.

He looked for signs of insects. The outer twigs of a Jeffrey pine had a sooty color; perhaps the coating was honeydew of the large brown Cinara aphids. The Jay flew to one of the boughs and found hundreds of the tree lice, their beaks in the wood of the twigs, feeding on sap. He ate aphids until he noticed a sawfly cutting into one of the pine needles with her sharp abdominal organ, to deposit an egg. She was a bigger beakful than an aphid.

As the Jay had alighted in the twigs, a chickadee had fluttered away. She knew she must give up the aphids if the Jay wanted them. He threw a "Tchah!" after her. A ruby-crowned kinglet and an Audubon's warbler had been picking over the needles. They too flew off, but rather as if blown by the breeze, not with such heartwarming acknowledgment of the Jay's dominance as the chickadee had shown. Sometimes one might profit by watching the insignificant smaller birds. The Jay heard the lisp of a Sierra creeper and discovered her, like an animated cocoon, hitching up the trunk. She kept shining the almost luminous white of her throat into the cracks, where there must be insects, for she often stopped. The Jay flew to the trunk, driving away the creeper.

He found flat-headed borers, both beetles and slugs. The green-bronze beetles were as vividly iridescent as the Jay. He got only two, but stripped off a loose piece of bark and uncovered four of the big moist grubs. He noticed that an adjoining fir had thin-looking sprays, flew over, and learned that the needles were being eaten by tussock moth caterpillars. The Jay ended the damage that five of the orange-striped worms were doing. He would return to this fir many times; without any tree-saving motive, he and his jay comrades would restore its health by the end of summer.

His hunger satisfied, the Jay relaxed on a shady bough until his energy was not needed for the work of his stomach. Immediately, then, there bounded up in him an impulse to make himself handsome. He returned to the spring, dipped in the water, and whipped his feathers until the dust was washed from quills, coverts, and down. Up on an oak branch, he whirred the moisture out of them. The vanes of a number were ripped. The Jay oiled his beak in the gland on his rump and pulled the beak down the feathers with a quick, sawing touch. The barbules locked together much as humans' zippers lock; the feathers became smooth and glossy again. Another whirring layered them.

With the grooming done, the Jay was a splash of magnificent blue, all his back and wings an unbroken blue surface that seemed more intense against the black of his head and breast. The blue was not true pigment as, for instance, a western tanager's yellow and scarlet are, but the prisms in the Jay's feathers made him look blue, more dazzling than if the blue had been real. His electric color actually was light, the reflected blue element of the daylight.

The bird's manner was equally brilliant now. He flicked about on the oak branch, snapped his wings, turned from side to side with motions dramatic and vivid. No hawk had a glance more dominant. Was the Jay really so sure of himself? Or was his confidence like the blue of his feathers, lost if a harsh touch broke the prisms?

The Jay started for the top of the tallest fir on the upper surface of Beetle Rock. That pinnacle was almost directly above the small meadow; since birds seldom ascend at angles of more than forty degrees, however, the Jay mounted the cliff in a spiral. He left the branch with a strong forward push into the canyon wind, swung around, and placed his wings well to the front, a position that made him tail-heavy and gave his body an upward tilt. He flapped vigorously, and also curved down the wings' trailing edges so that each had a deepened arch in which the wind caught and helped to lift him. The wind turned up sharply on striking the cliff; the Jay took advantage of the vertical current. At the top of the Rock he alighted on the fir's lowest bough and climbed by fluttering up from limb to limb—a ladder of more than fifty steps before he reached the banner-like tip of the tree, streaming eastward above the long, limber trunk.

From his green peak in the sky the Jay saw a golden eagle, like a mote of black dust, appear at the head of the canyon. It flew down, but over against the opposite ridge, for the Jay warned it to stay away from the Rock. He watched two red-tailed hawks at their nest on one of the Ash Peaks. They were likely to come out in the sky soon, and soar on the updraft. Let them ride it too close and he would send the great birds retreating—the Jay would, although his size was in the chickadee class compared with theirs. Around the edge of the Rock went his glance to make sure that no weasel or other predator crept up to surprise the animals warming themselves on top, on the granite now creamy with sunshine. A human being at the Jay's post would have needed binoculars to discover a weasel there below, but the Jay's eyes could see farther than a human's.

With a showy flutter his wings broke the sunlight. He tossed his head, and the brightness glinted along his beak. A cry seemed to explode in his throat. Ringing out over the Rock went the jay-scream that said, "Hunter!" Then—and what a fulfillment it was!—the deer pranced with alarm. The chipmunks and squirrels raced for hiding. The wingbeats in the trees vanished. An instant earlier all the mountain was stirring with life. The Jay called a warning and now every creature had fled or frozen. The Jay was master! Again he screamed.

But the scream broke off. What were those two mounds below on the Rock, as large as resting deer but without antlers or tails? They should be investigated. The Jay set his wings, drawn back to make him head-heavy, and launched himself from the branch. Down forty times the height of a human being he fell, then levelled out by advancing the wings and spreading and lowering his tail. The change in the shape of his body stalled him at the precise point to alight on a pine bough.

He slipped through the tree, observing the curious mounds from different angles. One mound broke open, arms flung off a blanket, and a man sat up. At the first move, the excited bird fairly tipped off his bough, but when he saw a human being emerge, his tail ceased to twitch and the gleam in his eye faded.

Humans were only starting to appear on the Rock in June. They were migrants like the deer and many of the birds. The Jay, who stayed through the winter, had seen many animals come and go, but all fitted, however briefly, into the Rock's wilderness ways—except the people. People were unpredictable. Every other species had a limited range of motions and sounds; chickarees climbed trees, for example, while ground squirrels did not; an eagle dove onto its victim, but a red-tail might chase a bird on the wing. One's whole security lay in knowing what any creature might do, but there was no safety with human beings, for they had no sure pattern of action; they might do anything. The Jay had a special reason for disliking them. They would not heed his voice, would neither frighten, like the prey animals, nor be driven away, like predators. The Jay's mastery of the Rock was not complete while human beings were there.

Yet they were interesting. The second mound had disclosed a woman. Now he would watch these two. They strolled out around the rim of the Rock, then up a trail to one of the cabins, with the Jay over their heads, never separating himself from the trees' shadows. At the cabin the

woman began to work at the stove. The man got a bag of peanuts and sat on a bench near the door. He tossed a nut to a robin.

The Jay dove from an overhead branch with his fine, particular knack, not gliding until his momentum was lost, but alighting with a strong forward speed, then bouncing gracefully to a stop. No robin made such a spirited landing. It deserved a nut, but the man didn't notice it.

The Jay hopped nearer, but not as close as the robin, who was receiving nuts as fast as he ate them. How did that bird, less adroit than the Jay in quick takeoffs, dare to go within reach of the man's hand and shoe? Back and forth at a safe distance jumped the Jay, as sprightly as a grasshopper. Each time he came down he flung his head so that his crest whipped from side to side. Amazing that the man failed to see him, or to hear him, for the Jay incessantly screeched. If the man would throw a nut only a little farther, the Jay would get it. He was frantic to have one, not only because he liked peanuts, but to keep up with the robin. Repeatedly the Jay bent his legs, meaning to match the robin's fearlessness, but his intention failed somehow and he always hopped sidewise instead of forward. He really was waiting for an instant when the man and robin would turn away and he could dart in and snatch a nut without any risk.

Here now came a pair of chipmunks, bounding with motions as fine as tiny foxes'.

"Oh, the darlings!" cried the woman.

She left her cooking to come to the man's side and help him throw nuts to both robin and chipmunks. The chipmunks sat up with the nuts in their forepaws, shelling and skinning them with their teeth. At last the woman, who threw less precisely, sent a nut far enough out so the Jay could get it. Faster than the speed of the nut, he jumped to it. He and a chipmunk reached it together. With his long beak, the Jay could have grasped it first, but the little chipmunk was bolder and took the nut. The Jay shrieked his disappointment.

Now the man saw him. He said:

"You get away from here. You're too noisy."

How humiliating was his tone!

"Go on, get. We don't like jays."

The man threw something. Could it be a nut? Eagerly the Jay hopped towards it. It was a piece of bark. The man scooped a handful of chips from the chopping block and began hurling them at the Jay. He strode out from the bench and the Jay flew into a tree. The man sailed the chips up at him and the Jay fluttered higher. He no longer screeched. From behind a fir spray he looked down and saw the man go back to his bench and resume the friendly gesture of giving food, so similar to the gesture of throwing in anger, yet so different.

But the Jay would yet have some of those nuts. He watched until the chipmunks began taking away the nuts to bury them. After a chipmunk had tucked his nut in the ground, whirled the earth over it, and bounded off, the Jay would uncover it. He cached the nuts in crevices of tree trunks, with bits of bark jammed above to conceal them. He ate only one, adeptly holding it on a branch with his foot while his beak broke off the shell.

When the peanut feast ended, he flew to his fir lookout. Now, however, he stayed in the deep shade of a middle bough. He perched with his blue feathers against the trunk, showing the world only the black front of himself. Into his retreat penetrated the screams of the Weasel and a squirrel, but the Jay was no longer inclined to lead a clamor against any hunter. He was a changed bird, so secretive that he seemed born and trained by habit for hiding. The predators could do as they pleased; the Jay's eye showed that he had a different, more personal interest in mischief-making.

He was so quiet that he was not even discovered by a red-breasted nuthatch searching for insects on the bark of the branch. The nuthatch flew down to its nest in a stump, shooting into the hole without alighting, a trick it had had to learn because it had smeared pitch on the hole to keep out ants. The picture of the nuthatch entering the hole dropped into the Jay's mind with peculiar sharpness. A wish formed around it. From his hidden perch he began to scan all the dead trees he could see.

He watched a large oblong hole in a stub, but with curiosity more than hope, since it was the home of that black and white giant, a pileated woodpecker. Some bird was tapping on a live tree; he? No, the blows came too fast. That would be the little white-headed woodpecker, smaller than the Jay. The Jay knew that one's nest, too. It was in the sky-horse, a twisted pine snag on the center of the Rock. The Jay held his eyes on its hole. He was about to fly down when the female's head appeared in the circle. The Jay settled back. On a fir stump he recognized the opening to a red-shafted flicker's nest, shaped wider than long by the broad-shouldered owner. A flicker was a large bird, so the Jay looked elsewhere on the stump. A chickadee flew from a knothole and the Jay tilted forward with eagerness. He could see a blurred movement inside.

The stump was below, at the side of the Rock. A wing-slide would have taken the Jay there in an instant, but he flew around through a chain of shadows among the trees.

Finally he reached a cedar bough near the stump. He saw the father chickadee come back to the hole, immediately followed by the mother. Then both parents left. They would be absent now for a while. The Jay could hear the cries of their young. In that hole was a meal that he would enjoy. He went and got it.

When the chickadee parents returned, the Jay was perched on the sky-horse, bare pedestal in the sunshine. He stood with legs erect, bill uptipped, tossing his crest. If he had seen any other animal rob the nest he would have screamed the fact to the neighborhood, but there was none to expose him. He called:

"Tchah! Tchah!"

It was not his warning. It sounded instead like a shout of restored defiance, as if he had gained back self-confidence lost to the man, the chipmunks, and the robin.

The chickadee meal was digested quickly. Then the Jay mounted the fir. He would station himself in the sky again, to protect the Rock's population from robbers and hunters, but now would fly a patrol, rather than perching.

He stood on the topmost branch until the wind surged full on his breast, then flung himself forward upon it. This was not a flight to take him anywhere, but only to stay aloft, so he checked all his large movements. By tilting his wings' undersurfaces forward, he could make his glide level. The wind was irregular, as disturbed as a river. The Jay rode its uneven stream without flapping, with only small, sinuous changes in his body's form. The shoulders, elbows, and wrists of his wings were ever slightly turning, bending, rotating, adjusting themselves to the air's pressure, pliant all over their surfaces. He steered with the wings even more than he did with his tail, though his tail, too, was supple, and helped the bird keep a steady course through a wind that would have tumbled him wildly if he had not made himself its responsive partner.

The beautiful flying ended, the Jay glided down to the ground. When he fluttered up the boughs of his fir again, he was singing, a melody humble and sweet.

White cumulus clouds were blocking out part of the Jay's sky. Soon they had shut away all its blue, and rain was filling the air. The Jay must retire to some sheltered nook. He was ready, anyway, for a rest. He perched in a cluster of fir sprays as quietly as if he were not aware of the lightning flashes, the shocks of thunder, and the rain falling so heavily outside the canopy of his tree. By the time the sun shone again, he was much refreshed. He, and other jays too, flew about in the sparkling boughs, often alighting as if they enjoyed dislodging the showers of drops.

A change could be felt in the weather. The breeze had died, and there was a new mellowness in the warmth. Under these conditions the vigilance of a jay would be very important. Already many prey creatures had emerged to relax on the Rock. The predators would resist the atmosphere's loosened tension, and, being familiar with the way of their quarry after a storm, would stalk them with sharpened eagerness. The Jay must protect the unwary ones.

The first hunter to arrive was a coral king snake. The Jay saw it gliding along a granite crevice towards two lizards, battling among dry, crackling oak leaves. The Jay flew to a bough over the lizards and cried his warning, but they failed to listen. The snake surprised them and captured one.

The Jay could discover no other predators, then, for some time. Yet the afternoon was still favorable for attacks. His anticipation built up and up, to an intensity almost unbearable. He watched the red-tailed hawks feeding the young at their nest. Each time they flew out over the canyon, he prepared for a magnificent outcry, but the hawks did not once approach the Rock. Finally the Jay could stand the suspense no longer. From the distant ridge he heard the high cry of the male hawk. He could imitate that cry—and he did, from a well-concealed perch. It caused havoc! As the Jay screamed the hawk-like threat, terror struck one creature after another. Every animal except the deer showed panic, and the Jay watched from behind a dense tangle of needles. This was a much more desperate disturbance than he could make with his jay-warning, which did not tell who the enemy was.

While the silence of fear still lay over the forest, the Jay slipped down into a clearing behind the trees and quietly searched for insects. When he was ready again to appear, he flew out—and saw an immense wing descending among the tree trunks north of the Rock. That was a hunter's movement, one with a plan in it never sensed in the stealth of prey. The Jay sailed into the grove. Ahead rose a great horned owl with a mouse in its claw. It flew up to a high sequoia bough, followed by the Jay, who sent out a summons that filled the air with approaching blue wings by the time the owl had alighted.

The owl was enclosed in a swarm of hovering, screeching jays. It looked at them from behind the yellow screen of its eyes. When the riot continued as if nothing ever would tire the jays, the owl gathered its dignity closer, and flew to a branch of a sugar pine. The Jay and his comrades strung behind. They would wear out this monumental patience. After some time the owl's eyes began to show flicks of rage, and then the clamor gained volume. At last the owl bristled its feathers, and with half-opened wings leaned forward in a threatening lunge. But it would not attack the jays, as they knew. Suddenly, still carrying the mouse, it dropped to a hollow snag and disappeared. The victory of its small tormenters was final.

The Jay started back towards the Rock, arrogance in his eye. On the way, his offspring discovered him, and burst forth from the pine where they had been perching, disconsolate. Their screams were indignant as well as hungry, but their father sped away from them, losing them in the intricate boughs of a cedar. He stole a mushroom from a chickaree's store as he passed its tree. And he stopped to taunt the man and woman who had humiliated him in the morning. They were standing on a trail, looking at the sky. The Jay perched above them, screeching that every animal should beware of them. As he went on to the Rock, then, all his movements were challenging. Near the edge of the granite he found the robin, his rival, drinking at a rain pool.

With a squawk that fairly scratched the air, the Jay dove to the robin's side. Approached so boldly, the robin surely would give up his place at the pool. But he did not! Instead he turned to the Jay with a vicious snap of his beak. Amazing little robin, a head shorter than the Jay; why was his bluff so alarming?

The Jay hopped up high in the air. That should intimidate the robin. But the robin hopped nearer, again clapping his beak. The two danced around each other, feathers on end. This certainly was to be a fight. But the Grouse stepped up to the pool now, and in her quiet way moved ahead of the others. Why did they not protest, and include her in their battle? The Grouse simply surpassed them in inner force. In sharing defeat, they forgot their anger.

The sun was close to the western ridge when the Jay returned to his lookout. He perched with no motions except the turn of his head and eyes. He saw the swallows cutting out over the canyon among the clouds of insects. Warblers fluttered around the trees for the insects in the branches. The Mule Deer Buck stood at the Rock's rim listening to the sounds of a day passing into those of a night. Directly below the Jay, a ground squirrel sat upright at the mouth of her burrow, forepaws folded upon her white breast. These creatures seemed to have forgotten fear. Should the Jay cry out his warning, and once more see them scatter?

If he was tempted, he did not do it. Perhaps he was satisfied to know that he could. So many events of this day had shown him to be a master that his ego did not need any further proving on June eighteenth. He flew down unaggressively to the ground for a late meal of ants. Afterwards he climbed his lookout fir to one of its thick lower branches, took a stand there, and let the lids close gradually over his eyes.

WHAT HAPPENED TO THE MULE DEER

OUT FROM THE TREES on one side of the meadow swept a great horned owl. Even in the deer's eyes, it was large. It passed low above their heads, a dark rush, gone before the air was quiet again upon the deer's faces. The owl was seeking prey, and the prey were there, mice all through the meadow. But the grass blades had protected them. The owl was gone, and this time no small life had been pressed out between its claws.

In the cabins behind Beetle Rock human beings slept, trusting the cabin walls to keep their food from bears. Animals that were food themselves also had let their alertness go, believing that in their burrows, trees, and rock-piles no predator could find them. The deer, out in their meadow, were still awake, watchful as they grazed. It was nearly midnight, but the deer, largest of prey creatures, had no dens where they could hide.

The Mule Deer Buck was nipping off new fronds of bracken near the meadow brook. One hoof stepped into the water. No other deer mistook the sound for a diving frog. Their seventeen heads turned towards the Buck, for he was the herd leader. If he was crossing the brook to leave the meadow, they would follow. But he stood quietly, facing the east, and the wind. After he swallowed his mouthful of bracken, he waited until the flavor no longer blurred the odors that he smelled. Then he moved the soft end of his nose up and down to sharpen the wind-borne scents. Tonight the wind was so strong that his nose would give him the quickest warning of any enemy approaching from the east side of the meadow.

The other deer were wary, too, but they relied on the herd Buck's greater caution. His ears lay back along his neck, reaching for danger signs from the west. The ears, more flexible than the ears of other deer species, turned to each minute sound, quivering to enfold it more completely. They responded to sounds that human ears could not catch, perhaps to a cave-in behind an insect's burrowing, to a mouse's panting, or milk in the throats of little flying squirrels. To be a leader, a deer must hear a predator's footfall while it was no louder than such innocent stirrings. The followers of the herd Buck knew that they could depend upon his listening, even though half his left ear had been torn away the previous winter, during the annual mating contests.

The deer were out in the meadow because they had learned where homeless creatures have the best chance of escaping predators—away from the trees on dark nights; in the deep shade under branches, but with openings for escape, when the moon is bright. There was no deer tradition to cover a night like this. The moon had not yet risen, but vapory small clouds had caught the approaching gleam and reflected it down on the grass. The leader could clearly see the herd's black tail-tips, tight against the white rumps. If he could see them, so could a cougar. Should the deer go back under the trees?

The speed of the clouds was wilder than any movement in the forest. No need for caution delayed the luminous spindrift, flying perhaps to a safer, freer world. The cloud-foam disentangled itself from the treetops on one side of the meadow, an instant later had blown beyond the treetops on the other side. It made this pit at the bottom of the foliage walls seem a little like a trap.

The Buck looked off into the tree trunks. There the night was dense, but not as black in a deer's eyes as in humans'. He watched for movements more than shapes, for the lifting paw or gliding head. What was the wave-like motion at the top of a log? The Buck's nerves bounded. But the sinuous looping was only the passing of the Weasel and her five kits. They were gone. A flying squirrel dropped from an overhead branch to the log, alighted in the weasels' frightening scent, and darted away. A gray squirrel, restless on this windy night, marched down a pine trunk. On the ground he stopped and called. He was a harmless creature, yet his voice had a tone of urgent cruelty. Now the wariness of the Buck seemed intensified. He leapt over the stream to leave the meadow.

He led the way northeast to an open grove where the deer spent moonlit nights. The younger deer would sleep beneath any convenient trees, but most of the older ones had favorite beds. The bed recognized as the herd Buck's was under a cedar's low branches. The tree was on a slight rise, so that the wind swept directly to the Buck's nose. And from there he could see into distant patches of moonlight that an enemy would cross in approaching the herd. It was a well-placed bed for a leader.

The Buck did not go to it at once. He stopped to browse on the leaves of a scrub oak near the cedar. The other deer, accepting his wisdom, were coming into the grove and finding their resting places. The order in which they came told something of their position in the herd.

First to follow the Buck was a forked-horn two-year-old, his satellite. After him came the eldest buck, whose weary weight swung from the framework of his shoulders. These two bedded down beneath a sugar pine next to the herd Buck's tree. Behind them a three-point buck, two four-pointers, and two does entered the grove and stopped at adjoining trees. Three does stayed in the deep grass of the meadow. The does would bear their fawns in a few days, and made all moves reluctantly.

The ears of the mature deer twitched with annoyance, now, at the approach of the yearlings. The six yearlings, tensely forlorn at being motherless, always stayed together and a little apart. They were as restless as birds fluttering on their perches; their hoofs plunged constantly, heads tossed, and ears jerked. They would call the attention of any predator to the presence of the other deer. Some of the yearlings were learning poise, however, and were working out their relationships to the herd. That one who struck so precisely with her forefeet would make herself boss of all the does within two years. But what good could await the anxious young doe, so nervously prancing, pounding back and forth in the meadow at scents and sounds she only imagined? The yearlings stopped beyond the others. Would it take them all the night to get themselves distributed?

One deer had not come. He was waiting until it would seem that he did not follow the herd Buck. There now was his confident step, and his lifted head, high and black and sharp against the light on the meadow grass. Tonight for the first time his antlers showed a new point rising in each front fork. No other buck in the herd was growing fifth points, though they were not uncommon among mule deer. They meant that this buck would have five tines instead of four, this year as last, and that the extra tines would give him a deadly advantage in the combat season.

The buck was a young outsider, just entering his prime. He had joined the herd during the mating contests of the previous winter. He had come closer than any other buck to defeating the present leader. Each time the leader and the new buck clasped their antlers and began the struggling pressure against each other, the stranger's fifth tines had come dangerously near the leader's ears. At first, by holding his head at a strained angle, the herd Buck had avoided the extra points. At that angle, however, he could not use the full strength in his neck muscles. Finally he had thrust his head forward recklessly, let one of the fifth tines tear through his left ear, and thereby recovered all his neck power. While blood flowed from the side of his head, he forced his weight against the challenger. Shadows lengthened before he felt the other give. Even then the stranger's weakening was so slight that the leader could not measure it. But he sensed it, and exhilaration renewed his energy. More quickly then, he drove the other back and down on his knees. The five-point buck had groaned submission, and the leader had relaxed, withdrawn his antlers, and stepped away. The does belonged to him still, and for another year the rights that went with his proved superiority.

Within a few weeks all the bucks had dropped their antlers and their antagonism towards each other. But the competition would arise again next fall. And again, it seemed, the outside rival would be armed unfairly. The leader must watch those fifth points grow, longer and longer, and finally lose their sheaths of velvet. Then he would see their owner polish them on willow bark. To win against them, he must well surpass the other's strength. But tonight the tines were only swellings on half-grown antlers.

Usually the five-point buck went directly to a bed near the yearlings. Now, however, he came to the oak where the leader browsed. He, too, began to nip at the leaf-buds, but he stayed well out of the leader's way.

The swing of the earth had taken the deer's grove to the point farthest from the sun. It was midnight in time; was the Buck aware of time? Perhaps he was more aware of it than human beings are. Hours and weeks are not measured in the wilderness, but the life there is determined by the sun's days and the moon's months even more than is the life in cities. The Buck must also adapt himself to nature's smaller movements, to winds, clouds, the lengthening of shadows, leaves' growth, the rise and fall of streams. Being dependent on such delicate transitions, might he not sense a change when the ground beneath him curved no longer away from the sun, but towards it, when his night of heightened danger began to end?

And since his place in the herd was related to time's changes in himself, might he even be aware of the important turnings in his life span?

At midnight the Buck walked away from the oak leaves, and farther out into the grove. He may only have been nervous. For more than seven days he had not seen or scented the Beetle Rock cougar. The cougar hunted over a wide route, returning approximately once a week. The Buck had had ten years in which to gain a feeling for the interval between its visits. When it was due, the Buck was restless. He drained the wind, and started at every falling twig, at every stirring of a bat or mouse.

Finally he was satisfied that no predator stalked in the grove, and turned back towards his cedar. The moon was up now. As he passed from tree to tree, he crossed its light, which touched to a blaze the shine that smoulders only in wild eyes—red-white in the Buck's eyes.

The five-point buck left the oak and he, too, walked into a channel of brightness. But why is he not going to his bed near the yearlings? Incredibly, he approaches the leader's tree. He is pawing at the leader's bed. Finding no rocks or branches, he drops a knee to lie upon the needles.

The leader's anger was a hot, dark flood. He reared and struck at the other with his forefeet. The five-point buck leapt up and towards him. The night was violent with the stomping of their hoofs and with their snorts and hisses.

The other deer all rose from their beds. Perhaps a sense of their presence, of the startled herd around them, came to the five-point buck with the pressure of a custom. He stepped back, away from the leader, withdrawing his untimely challenge. Briefly the two deer faced each other with no motion but their heavy breathing, then the leader turned and lay upon his bed. After the five-point buck had stood where he was a little longer, he pawed out a bed for himself beneath the same tree, but with the trunk between him and the herd Buck. The leader chewed his cud, showing no sign that he was aware of the other. Both deer kept their heads up; both were alert, defensive in their resting.

Through the rest of the night the herd stayed under the trees. They were still forms, with their tension gathered deeper in them now. From time to time the leader closed his eyes, but his ears and nose were informing his sleepless nerves.

The weight of the oldest buck lay on the ground more flatly, his quietness more inert, than the other deer's. When the dawn first marked him, his eyes were open, but he seemed remote from the crystal chirping of the birds, from the chipmunks' waking energy, and other creatures' lively return from sleep.

He could show the pattern of a deer's aging to the herd Buck, watching him. The old one's shoulders were high between his tired back and his tired head. His legs were stretched out straight to ease the cramp in his knees. On his hind parts, his thick gray winter fur had not been replaced by the summer coat of rusty tan. His antlers had not yet grown to the first fork; they were forming only one fork these years. In June the antlers of all bucks ached at the bases and were sensitive in their turgid sheaths, but the old one's may have been especially painful. For that reason or some other, his temper was quick to flare.

No muscle sagged in the sides of the present leader. His back was one smooth, curving line from tail to antlers. He lay with his legs all folded beneath him, so that a single straightening move would spring him farther than any predator's pounce. Yet this morning his tail twitched and his ears were nearly as wild as a fawn's. His composure had been upset by the presence of his arrogant rival, and perhaps by the strain of the windy night, with the cougar expected. When he left his bed to go out in the meadow, his faithful forked-horn follower rose to go with him. Irritably the leader struck

at him. The forked-horn watched with astonished eyes as the leader walked away. Then he wandered off alone among the trees.

Now in the dawn there were rounder curves on the tree trunks. Countless details which darkness had covered were coming back into the meadow and the forest. The snow plants and mushrooms were back on the ground, the cobwebs were on the ferns, the rippled sand was in the brook, and the silken shadow below the layer of flower heads. Boulders were clasped in the upturned roots of the sequoia log.

The Buck found a patch of Spanish lotus. Its puckery freshness was the taste he always wanted in his mouth in the early morning. But on this day he ate little of the lotus. He was restless, and would finish his first meal somewhere else. Soon the other deer would be leaving, too. In pairs or singly they would start away on trails to groves, to thickets, and smaller meadows. They would not be a herd again till nightfall, though various ones would meet where they were going.

The Mule Deer Buck turned east towards Beetle Rock. There, beneath the fir that overlooked the granite field, he joined a group of animals every afternoon. Most of them lived around the tree; they knew each other and had worked out peaceful ways to share the spaces of shade and sunlight, the food, the water at the nearest spring. In a way, the Buck was also a leader among them, for they relied upon his watchfulness.

All the way to the Rock he found animals that he knew. Along the stream was a separate neighborhood of those who liked the shelter of the gorge. Descending the steep trail, the Buck saw first a pair of gray squirrels. This morning the squirrels were absorbed in a grave game, a chase more like a dance. Beneath a scrub oak they pursued each other in small intricate circles, first the female leading, then the male. Above them floated their long tails, swaying with elaborate grace. The patter of their feet was their only sound.

The birds were leaving their night roosts, finding their seeds and insects, starting the busy play of fluttering from one side of the stream to the other. That weaving of their flight across the water's flow would fascinate them all day. The Buck approached a red-naped sapsucker drilling for his drink on a pine trunk. As the Buck passed, his antlers could have brushed the bird, but they frightened it no more than a blowing branch would. Down on the edge of the trail a fox sparrow scratched for food, with forward jumps and backward scrapes of its feet. It didn't miss a jump when the Buck walked by, for it knew the caution of deer's feet, how they avoided stepping on small creatures, even on flowers.

Finally the Buck leapt across to the other side of the stream, came out on an apron of sand and put his mouth into the water. The stream was quiet here. Down in its brown, wet clarity he could see a bird, a dipper, pushing itself along with feet and swimming wings as it searched for larvae under a sunken branch. The dipper's head bobbed forward, then the bird made an upward dive from the water, and carried the food to its fledgling, who stood near on a wet rock. The mother, too, perched briefly on the rock before she went into the stream again, and the two gray birds ducked up and down in time with the ripples.

The diagonal paddles of the water skaters were snapping together energetically, sending the watery-textured insects darting over the surface. Six minnows, who had been resting under a stone, glided back into the swiftest current and took their stand against its pressure. All day they would fight the stream to make themselves strong.

And now, with a ruffling whir of wings, the wounded Grouse arrived. The Buck expected her, but she vanished after she alighted. The dipper had time to find two more larvae before the Grouse appeared again. Where had she been? Perhaps in plain sight. No other creature had so unearthly a talent for making herself invisible, by her markings and by the stillness that seemed to remove her altogether. The Buck finally found her when he heard her beak click over a beetle. Slowly, with a sort of waiting awareness, she walked along the sand and picked up other insects.

The green sedges near the water were sharp in deer's mouths. The Buck preferred the tender grasses higher on the stream bank. He and the Grouse climbed up to them with movements so harmonious that they seemed to show a silent communication between these unlike creatures. The two stayed near each other, the Buck searching for delicate new blades of grass, the Grouse for worms.

As he grazed, the Buck was listening to the confident sounds here, feet running and bounding, wings whipping into flight, voices calling, chattering, singing. The streamside animals had a fearless way, perhaps because the gorge was a tangle of natural screens for hiding—of leaves, logs, rocks, and piles of flood wreckage. The cover was not the kind a deer could use, but the Buck seemed soothed by the lack of strain around him. More spring came into his movements, and even a little playfulness.

The squeak of a shrew was like a prick of threat. Smallest of mammals, but one of the cruelest, she slipped along past the Buck's nose, leaving an acrid scent. He raised his head and sensed that the squirrels' feet had stopped tapping, too abruptly. Why? His legs moved tensely; there was alarm in his nerves. Cougar? The birds still sang; a chipmunk cracked an acorn. The Buck's fear eased. His mouth went down for more grass. But when he passed beyond a boulder into a flow of wind, he found a new scent, sharp as a scream. It was not the scent of cougar, but of its relative, wildcat.

Quicker than thought could have sent him, the Buck was up the bank, was clearing logs and rocks with bounds three times his length, up over lilac and hazel bushes, up past the humans' trail, up the gully of a dry creek, up an animal path now, up and up and up. The effort was excessive, but the Buck was fleeing from his nervous fear of the cougar more than from the cat.

High on the slope he stopped and whirled to face the trail. Had the cat pursued him? He waited, eyes, ears, nostrils tense. His bounding had caused a startled silence here, but soon a chickaree and a jay returned to their foraging. The Buck relaxed, and when he did, found that his legs were shaking uncontrollably. He felt that he was smothering for lack of breath. This was the first time any chase had winded him completely. He lay beneath an oak.

Now another buck was bounding up the trail. It was the forked-horn, but he was not in flight; when he stopped he did not turn to face the gorge. He knew nothing of the wildcat. He had seen his comrade pass and had come to meet him, leaping with the exuberance of a young deer.

The herd Buck seemed glad that he had come. The two deer walked along the slope towards Beetle Rock and the dogwood thicket where they often spent the morning. There they would rest, enclosed with chinquapin brush, and chew their cuds. When they reached the thicket, the herd Buck ate for a while at the pungent leaves of the chinquapin. His interrupted feedings had not yet given him twelve pounds of green food, which he needed for a meal.

Finally he let himself down near the forked-horn. Each knew an opening through which he would escape if a predator entered the brush; they were enjoying the only ease a deer knows, ease based on readiness. Or the forked-horn was. The herd Buck did not feel even that security. A mouthful at a time, he brought up the food he had ground imperfectly and chewed it finer, meanwhile waiting in his muscles and his nerves for the return of his full strength. He still sensed an unfamiliar weakness from the exertion of his uphill flight.

His antlers throbbed from the increase in his blood pressure. Twice he laid one fork on the ground for the comfort of its touch. Biting deer flies clung to his nose; they had beaded it with blood. Several new wood ticks had fastened upon him. He bit off those he could reach and was going to get the forked-horn's help with the others when a rising storm distracted him.

The storm was stimulating. It broke quickly and with violence. The rain seemed not to fall, but to be hurled upon the forest. Thunder and lightning filled the sky with the power of wild destruction. The Buck and his young companion moved beneath a cedar, where they stood and watched, comparatively dry and much excited.

The thunder and lightning departed before the rain did. For some time the rain continued at shower-strength, an easy-falling freshness that drew a spicy fragrance from the leaves and a good, damp, woody odor from the bark. Dust, granite, even the dead leaves on the ground, smelled clean. The Buck felt ready again for any encounter. The storm had renewed his energy, as rest had not.

He saw that there were ticks on the forked-horn's neck and chewed them off for him. The forked-horn removed the ones on the leader's back. Each was exceedingly careful not to hit his antlers on the other's, a precaution purely in self-interest, but there also was a kind of tact in the motions of the deer around each other, a graceful giving way, a lack of wilfulness.

Without warning of scent or sound the Coyote came around the brush. The herd Buck only glimpsed the humped speed of the enemy flying towards them. He sprang back, body low, ready for a great leap, but he waited. He let the young buck bound away, a decoy, drawing the Coyote out of sight. Then the leader trotted off in the opposite direction, noiselessly, almost as slyly as a cat. He never before had reacted thus to an attack. In his early years, the Buck had been tricked like that by older deer to save their energy. He had forgotten the strategy, but his nerves remembered it. Now he had a new defense technique, adapted to some new stage in his physical progress.

On the lower slope the Buck stopped in a copse of young pines. The deer trail from the stream passed by the copse. Up the trail was climbing one of the Buck's does, a two-year-old, soon to bear her first fawn. Lost were her own fawn traits, the intensely questioning eyes, the secretiveness, the distrust. When she saw the Buck, she joined him. Standing beneath the little trees, she reached up and back with her mouth to eat some of the needle-buds. The needles were strung with raindrops, each one catching the sunshine now. The Buck browsed on the needles near her.

The shadow of his antlers lay upon the creamy fur of her full side. But the antlers' shadow was blotted out by a larger one. The five-point buck, who had been following the doe, walked between her and the leader. He too began to browse.

At this season the leader was not usually jealous when other bucks approached the does. He had had them all for himself in the winter, when they were important personally. This day, however, there seemed something more than amorous interest in the way the five-point buck had cut between him and the doe. Perhaps the manner of the five-point buck was always challenging. At least, the herd Buck felt himself displaced. The hair on his back rose, and his breath came quick and wild. Just then, however, the doe took fright at a falling cone, and shied, and raced away. The five-point buck turned from the pine trees to a fir with staghorn lichen on its trunk. The lichen was browsed off to the height of most deer's mouths, but the five-point buck could reach a little higher. He chewed at the lichen, apparently giving all his attention to keeping his antlers away from the tree.

The leader's annoyance was not dispelled. But he started on to the fir tree at the Rock. Soon he had lost sight of the five-point buck, yet his tail flicked, his ears whirled, and his hair still bristled.

At the fir he found a disturbance that increased his agitation. The forked-horn buck had disappeared, but the Coyote was there, digging into the Weasel's den across the draw. The Weasel was shrieking, jays were squawking overhead, and a ground squirrel piped a shrill whistle. The Buck stood under the fir, his nerves distraught. At his feet the helpless little Deer Mouse bounded from leaf to twig in a search for some small refuge.

When the Coyote finished the weasel meal, the Buck was ready to break into flight, but the predator turned off down the Rock. The sentries became silent. All the animals remained cautious for a while, then one by one they dared again to flutter, to leap, to chirp.

The anger of the Buck had still no chance to subside. Out of the thicket behind the tree came the nervous yearling from his own herd, her poise completely scattered. Wherever the Buck turned, she was in front of him, jerking up her feet and flinging her head above a swollen throat. The Buck tried to browse on the soft new needles of the lower fir boughs, ignoring the doe. But she pranced so near him that he was forced to jump aside, and his antlers crashed against the branch.

At the staggering pain, his fury broke. He struck at the doe, and the pointed hoof of his forefoot tore into her shoulder. She bounded across the draw and was away, flying with uncontrolled speed, senselessly turning the way the Coyote had gone.

The Buck stood under the fir, trembling, waiting for peace to come back into his legs. He was scraping his hoof through the fallen needles to clear a bed for himself when he heard a rustle in the draw. Perhaps a lizard or chipmunk was stirring the dead leaves, but the Buck must be sure. He held his eyes on the leaves. Out of them slid the long bright body of a coral king snake.

The Buck was never a predator, but now he stood as finely tense as any stalking cougar or coyote. He saw the snake glide up the gully-side towards the fir. Its motion became slower. Under the edge of the tree it stopped. Perhaps the warm earth here seemed a pleasant bed on which to lie and digest its food. But it had intruded upon the resting place of the Buck. The Buck watched without a movement, almost without breath, as the snake began to coil itself.

He waited until the snake had drawn up more compactly, and then he moved forward with an arched, precise step. He began to circle the snake. It had seen him; possibly it sensed its danger but knew that it could not escape. With a terrible deliberate fury, all the accumulated fury of the day, the Buck moved on, around and around. His neck had a forward thrust, his hair was up, and his eyes were hot with a cruelty that seldom swells into a deer's eyes.

A ground squirrel, a chipmunk, and various birds watched, tense and silent. Three jays in the fir were too fascinated to scream. Still as slowly around the snake moved the Buck. The snake's head turned as the Buck passed in front of it. Its scarlet tongue darted in and out of its jaws; its eyes glittered. Around again and still again went the Buck. Then he was high in the air. His four sharp hoofs together, he brought his full weight down upon the snake. Many times he leapt and struck it. The snake was broken flesh, then it was pulp, and then was only a stain upon the ground. Not until all his strength was spent did the Buck cease the attack.

Never before had he been so exhausted as now, and never so unnerved. If a coyote or cougar had attacked him, he would have been able neither to flee nor to defend himself. He saw the Grouse walk with her beautiful balanced step into mottled sunshine, and let herself down upon the needles. A breeze blew over her and her feathers rose, the after-shafts filling the spaces between with down. She was round and soft, and looked completely comfortable. Along the line of the Rock against the canyon, her eyes found some reason for peace.

The Buck knelt and lay near her upon the weathered needles. His breath was short, and his heartbeat was like the pound of hoofs. He brought up a cud of chinquapin leaves. He could only wait, hoping his strength and poise would return.

Around him were various creatures that he knew as well as he knew his herd—little companions who never competed with him, and never feared him, since he had no taste for flesh. They went about under his very nose and hoofs, absorbed in their own doings but communicating their moods to each other, and possibly even to the Buck.

He watched a pair of chipmunks, two small, separate lives between which there had sprung the tension of love. This afternoon they were in the full play of it, stretching out the elastic invisible cord, letting it snap, stretching it, snapping it.

Their play was a chase, none ever faster, lighter, or with pauses more sweet. The male waits, his back to the fir, fore-paws clasped to his breast. The female watches him from the trunk above, where she hides in the bark. She steals down, touches his head with her nose, and he turns. He reaches up and their faces stroke each other, cheeks and noses stroking with motions facile and tender. She is gone, and he after her, up and around the tree, and down. In front of the Buck he meets her, and the two tumble together so fast that even the Buck's quick eye cannot disentangle them. Then they speed away into the brush. They will come back. All afternoon, all summer, the game will continue, even after their young begin to peek out from the nest hole. The original two, seeming much too ingenuous to be parents, still will play. Neither will drop his end of the fine, resilient cord.

The Buck drew a breath that deeply expanded his sides, as if he had recovered some delicate tension from the chip-munks' example.

Another small comrade, a golden-mantled squirrel, was walking around him on hind legs, eating the seeds of flowers and grasses. He pulled down the tops of the taller plants with forepaws as deft as hands, and stood up nibbling the seeds while his eyes met the Buck's with confidence equal to his. The furry gold button that was his nose began an interested quivering. Apparently he smelled the grass juice of the Buck's cud, for he walked close to the Deer's face, a face as large as he was, and reached up with his paws. From the distance of his different nature, the Buck looked down at the squirrel, and continued chewing. The squirrel did not often seek food from deer, but it was typical of him to override habit, and to risk such an overture. All his ways had a stocky stamina, possibly contagious.

The Grouse, rested now, moved farther into the sun and began a dust bath. The Buck became conscious of a new discomfort, a pain he never had felt before. It was in his knees. Perhaps the storm had cramped them. Would he dare to unbend his legs, to lie with them outstretched? He tried the new position, and the relief was like food in a hungry mouth. He would not be able to spring up quickly if an enemy appeared, but danger seemed to have lost some of its urgency. The Buck was calm again, but listless.

In the cherry bush nearby was a pretty liveliness that might be stimulating. A calliope hummer spun from bloom to bloom, cerise rays darting from its throat. A junco declared from the outermost twig that his territory extended to *here*. His song trilled as brightly as if the vibrations of light had been made audible, a pleasant way to assert one's rights. And down the fir tree flashed the Chickaree now, to tease the Grouse. This was his daily recreation, rousing to watch. While he sputtered around the bird with saucy taunts, she continued her dust bath, even repeated it, until one of his squeals finally exploded her anger. Out flew her feathers then, and she danced and hissed. The Chickaree climbed to the fir-top, but was down again quickly to show her that she could not intimidate him either. Fiery little tease—no one could watch him and remain inert. And the Steller's Jay, another tease, added some excitement by imitating a red-tailed hawk.

Once more the Buck was a deer to lead deer. His eyes were perceiving smaller movements, and his ears were flicking to slighter sounds than most deer would catch. He had a leader's extra margin of alertness again.

He saw the forked-horn coming up the draw and felt ready to get up and join him. In late afternoon the deer went out to browse on the Rock itself. The Buck had only to cross the draw, but that creek bed was a litter of boulders and logs, a hazard to all but the steadiest feet. The Buck rose awkwardly from his new outstretched position and started into the granite wreckage. He walked as if he were not sure what his feet would do. One foot did step on a stone that tipped, and the Buck went down on his knees. The only harm done was to his confidence. Rest and the example of his spirited companions had healed his nerves, but now he seemed to sense that no example of energy could give strength to muscles or power to a heartbeat.

Out on the Rock were half the Buck's herd—the yearlings, two does, a four-pointer, and the five-point buck. The heads of the deer were buried in the brush; each was concentrating on his effort to gather, leaf by leaf, the great amount of foliage required to nourish a grown deer. The leader and the forked-horn browsed along the north side of the Rock, above the draw.

The morning storm had cracked a dead branch that overhung the Rock from a sugar pine. Now suddenly it broke and came crashing down through the lower boughs. All the heads were quickly out of the brush and the deer were moving towards the pine, for the branch was covered with staghorn lichen, their favorite food. They stopped, however, when they saw it falling near the leader. Since he was the herd boss, it was his of course if he wanted it, and he did.

The branch hit the granite and bounced his way. He stepped towards it. But the five-point buck leapt forward, ahead of him. The leader's hoofs shot up. The five-point buck will compete for this lichen! Both deer have reared. The five-pointer fairly boxes with his forefeet, a steady attack. His blows are swifter and more precise than the exhausted leader's. One hoof has struck the leader's chest. Its thud can be heard above the bucks' hoarse hissing and the ring of their hind feet on the stone. The leader flings back his head in a gasp. Then his forefeet come to earth gropingly. He stands bewildered as the five-point buck takes possession of the prize.

While the long branch was cleaned of lichen, the Mule Deer remained motionless, recovering his breath, feeling himself transformed into a buck who could not defend his rights when they were challenged.

The sun was low. Among the darkening mountain ranges, Beetle Rock held up a late warm island of daylight. Birds skimmed and chipmunks bounded freely, as if no hunter ever would enter an atmosphere so undefended. The Lizard, returning to his den, looked up at the Deer so strangely quiet. The movement of the little creature caught the Buck's eye; he turned his head to watch it. Then he walked a few steps and began to browse.

All the deer were foraging near the draw now, working towards the trail they would follow down the slope and across the stream to the meadow. Two of the yearlings started into the draw. But they stopped, for they saw that the Buck was alarmed. His head was up, his body tense. He had heard a bear cub call, and he was the only deer who did hear it.

With the secretness of long skill, he shifted up and down the Rock, sharpening all his senses for more bear signs. No other deer knew why he was startled, but they would not move as long as he was nervous. Since his ears were pointing across the draw, they listened in the same direction, but they heard nothing.

Then clearly to them all came the breaking of bark under the Bear's claws. Soon afterwards her scent was strong on the Rock, for she and the cubs were walking towards the draw. The yearlings bounded back towards the rim, but the Buck's experience with his enemies told him when he was safe as when he was in danger; he knew that bears never would expose themselves on Beetle Rock at sunset. The bears remained among the trees on the other side of the draw, the Buck following their progress by what he heard and smelled of them. Finally he knew that they were out of the neighborhood.

The other deer continued to watch the Buck. His actions showed them the stages of his returning confidence. When they saw that he had forgotten the bears, they felt reassured enough to start for the meadow. One by one they disappeared between two granite blocks, and into the draw.

By the time full darkness lay in the meadow, the deer all would have assembled there. They would turn their eyes towards the Buck whenever their heads were lifted from the grass. They would know if he stepped in the brook; they would wait for him to guide them out of the meadow when the moon rose. On this June eighteenth, physical dominance in the herd had passed to the five-point stranger. But leadership in wisdom would belong to the Buck for as many years as he deserved it.

The Buck had his own way of ending a day. He liked to stand at the edge of the Rock while the sun moved towards the ridge across the canyon. After it was gone, he turned out over the Rock's rim and down a slanting ledge to the shelf below on the cliff, and thence to the trail. Tonight, as usual, the forked-horn returned ahead of him to the meadow.

At this time sounds were very clear, as if the air resisted them no longer. The breeze, brushing up the wooded canyon-sides, found corners and blades of granite at the cliff, and curled upon their edges with a resonant hum. The river below was a colder murmur, waters leaving the canyon, rushing away.

The animal sounds were the ones that touched the Buck's ears into motion. They told him that one company was turning home for sleep. The Chickaree's claws were catching into the bark as he climbed the fir to his nest; the grouse cock beat a late call; a golden-mantled squirrel barked to summon its young; last mouthfuls of leaves were being pulled from branches. Less tranquil, on the trail below, a boy's excited voice cried out.

The Buck stood tall, his legs straight and his head lifted. He was still except for his ears. Sometimes they listened separately, one curving forward under his antler, the other one back, point quivering towards the trees. Then the ears would swing to the same sound, as smoothly and subtly as if they were hinged together.

Other sounds meant that a different company had finished their sleep and were setting out for their waking hours. Distant robins warned that a bear approached. The down of an owl's wings ruffled the air. The flight of uncounted insects blended into a fleshless drone; as a bat pursued them, its snapping mandibles clicked. One voice wished and dared to sing. From a cranny at the Buck's feet came notes like raindrops hanging from the tips of leaves. They seemed to belong with the voices at the stream in the morning, lifted without danger in the daylight. This one, bravely raised in darkness, was the voice of a singing deer mouse.

Suddenly the Buck sprang back, his legs aslant, tensed, ready to bound away. For on the granite shelf below, he saw the cougar. It was pacing eastward. Even when it reached the outer corner, where the cliff broke into a talus slide, the cougar continued swiftly, obviously not hunting. It was starting away from Beetle Rock, from this part of the mountain, where it doubtless had spent the day. It would be gone again for a while, as the Buck knew. Its long tail shrank, then vanished, around an edge of granite. When it vanished, the principal enemy of the deer was gone.

Now there was a change in the dark shape of the Buck above the canyon. His head was a slightly falling, not a lifted line. A trace of swing came into his back; his knees showed the beginning of a bend. The cougar's departure had lifted much of his fear, but this Deer was resting from the strains of more than one day.

For ten years, he had not relaxed completely—his muscles often, but never his alertness. He had poised himself on the forest movements, scents, and sounds as tirelessly as if he had been a bird, born in the air, who must soar through all its life. But finally he would be able to alight. Did he know that now?

APPENDIX
CAST OF CHARACTERS

LONG-TAILED WEASEL: *Mustela frenata*
From about eight to eleven inches long (the male is larger and weighs about twice as much as the female), with long, thin, furred tails about half the length of their bodies, long-tailed weasels of the Sierra and the Southwest are brown on their backs and the outsides of their legs, with cream or tan bellies and throats. Good swimmers and climbers, these intense, sinuous predators are versatile, living in forested, brushy, and open areas, preferably near water. They range from Canada to Bolivia and live throughout most of the United States.

SIERRA GROUSE (BLUE GROUSE): *Dendragapus obscurus*
The male blue grouse is a dusky or bluish gray, about fifteen to twenty-one inches long and weighing as much as three pounds. The booming hoots of a courting male can sometimes be heard from a quarter-mile away. As he hoots, white feathers on both sides of his neck spread to reveal a bright yellow air sac, and as part of this courtship display, the red combs over his eyes stand up, and his tail fans over his back. Females and their young are mottled brown with dark tails.

Resident from southeastern Alaska and the Northwest Territories to California, Arizona, Colorado, and New Mexico, blue grouse live in burned areas, the brush of coastal rain forests, coniferous forests, subalpine forest clearings, and slash.

Chickaree (Douglas's Squirrel, Pine Squirrel): *Tamiasciurus douglasii*

"Though only a few inches long, so intense is his fiery vigor and restlessness, he stirs every grove with wild life, and makes himself more important than even the huge bears that shuffle through the tangled underbrush beneath him."

—John Muir, *The Mountains of California*

One reason that chickarees are important is that their huge stores of seeds and cones are a food source for many other creatures. Chickarees are from about ten to fourteen inches long with four- to six-inch tails, and they weigh from five to ten ounces. Their upperparts are reddish or brownish gray, changing to a chestnut shade at the middle of the back and onto the tail, except that the last third of the tail looks black. Underparts are gray to orange. There is a dark stripe on each side in summer. In winter the stripe disappears and the

chickaree's overall color is grayer. The underside of the long, black, bushy tail is rust-colored in the center, bordered by a broad black band with a whitish edge. Chickarees live mostly in coniferous forests from southwestern British Columbia into California.

Black Bear: *Ursus americanus*
The black bear and the grizzly (extinct in California since the 1920s) are California's only native bears. Still found in much of the United States, *Ursus americanus* is sometimes black but often brown. Males on all fours can be more than three feet tall; standing upright, five feet. Females are much smaller. Their diet includes grass, bulbs, berries, nuts, and other plant foods, insects and honeycombs, fish, carrion, and occasional small mammals. Powerful swimmers and competent climbers, they can run as fast as thirty miles per hour. Most black bears winter in dens, where they can live for up to five months without eating. Cubs (usually two) are born in January and February after about seven and a half months' gestation.

Western Fence Lizard (Blue Belly): *Sceloporus occidentalis*
The western fence lizard favors rocky and mixed-forest areas from sea level to above nine thousand feet but adapts to a wide variety of other conditions, frequenting stone fences, fence posts, and old buildings as well. Six to nine inches long, these lizards are olive, brown, or black with

dark blotches on their backs and down their tails. Males have bright blue bellies, and the undersides of their legs are yellow. Adult males have blue patches on their throats. The entire belly and throat of the adult male Sierra fence lizard (*S. o. taylori*) are blue. Western fence lizards communicate by bobbing their heads and flattening their sides to show off their blue patches, probably to attract females and drive off intruders. They are diurnal and easy to find, even in winter on mild days.

Coyote: *Canis latrans*

The coyote's common name comes from "coyotl," the name that Mexico's Nahuatl Indians used for the animal. The scientific name means "barking dog." The coyote's most distinctive vocalization is a series of barks and yelps followed by howling and ending with short yaps. Members of a band use this call to alert others to their locations, and one call usually leads to a chorus.

Gray or orange-gray above with buff underparts, prominent ears, long legs, and a bushy tail with a black tip, coyotes weigh from twenty to forty pounds, and sometimes as much as fifty-five pounds. In the West they live mostly in open plains; in the East, brushy areas. They normally run at about twenty-five to thirty miles per hour—sometimes as fast as forty miles per hour—and they can make fourteen-foot leaps. Despite years of being trapped, shot, and poisoned, these crafty and adaptable animals have maintained their numbers in the western United States and are on the rise in the East.

DEER MOUSE: *Peromyscus maniculatus*
This tiny creature is about the same size as a house mouse—from about four and a half to eight inches in length, including its long tail. Its upperparts are gray-brown to reddish brown, and its underparts are light or white. Its short-haired tail is dark on top and light below. Deer mice live in habitats ranging from prairie to brush to woodland, from Alaska to Mexico. They feed on seeds and nuts, small fruits and berries, caterpillars, and other insects, storing food for winter in hollow logs and other caches. The deer mouse itself is a readily available and hence important food source for many predators.

STELLER'S JAY: *Cyanocitta stelleri*
Named for German naturalist George Wilhelm Steller, this easily recognizable bird is the only western jay with a crest. The color of the Steller's jay moves from charcoal black to blue-gray (vivid blue in the right light), top to bottom. The Steller's jay's call—inevitably described as grating, raucous, or rasping—can be heard from southern Alaska to Central America in forests, stands, and groves of conifers.

MULE DEER: *Odocoileus hemionus*
This is a medium-sized deer, about three feet tall, with a stocky body, long, sturdy legs, and a stiff-legged, bounding gait. Mule deer are distinguished by their large ears, which move independently and almost constantly, like mules' ears. They range from the southern Yukon and the Northwest Territories south through the western United States to Wisconsin and west Texas. The mule deer of the Sierra Nevada has a white rump extending above its white, black-tipped tail. They are active day and night, year-round, but are mostly nocturnal in areas where they are hunted.

ABOUT THE AUTHOR

AFTER SPENDING MUCH of her childhood studying nature on her grandparents' land in Ohio, Sally Carrighar (1898–1985) went on to travel in northern Michigan and the Rockies, learned how to track in the Canadian woods, and worked as a guide at a fishing lodge in the Ozarks. During the 1920s and 1930s she wrote for motion picture companies and for radio, but *One Day on Beetle Rock*, originally published in 1937, was her first venture into nature writing. A popular book, it was also well received by critics. Several of her later books take place in the Arctic, where she spent nine years after receiving a Guggenheim Fellowship to do fieldwork there. In addition to writing numerous magazine articles, Carrighar authored ten other books, including a play, a novel, and her autobiography, *Home to the Wilderness* (1973).

DAVID RAINS WALLACE has published over a dozen books on natural history and conservation, most recently the *Official National Park Handbook* for Yellowstone. His third book, *The Klamath Knot*, won the John Burroughs Medal for Literature in 1984 and will be reissued by the University of California Press in 2003.

CARL DENNIS BUELL is an illustrator and naturalist whose work has been featured in museums and zoos throughout the country, as well as in numerous national magazines. He has been clawed, pecked, or bitten by most of the species of wildlife featured in this book. He now lives in upstate New York with the world's greatest dog.

OTHER CALIFORNIA LEGACY BOOKS

Death Valley in '49

By William Lewis Manly, edited by LeRoy and Jean Johnson, introduction by Patricia Nelson Limerick

Originally published in 1894, this California classic provides a rare and personal glimpse into westward migration and the struggle to survive the journey for greater opportunity.

Eldorado: Adventures in the Path of Empire

By Bayard Taylor, introduction by James D. Houston, afterword by Roger Kahn

A reprint of one of the most valuable firsthand accounts of the California gold rush, *Eldorado* is the quintessential recounting of this era, as seen through the eyes of a New York reporter.

Fool's Paradise: A Carey McWilliams Reader

Foreword by Wilson Carey McWilliams, introduction by Gray Brechin

A collection of over twenty-five essays that span nearly three decades, *Fool's Paradise* examines some of Carey McWilliams's most incisive writing on California and his home city of Los Angeles.

Lands of Promise and Despair: Chronicles of Early California, 1535–1846

Edited by Rose Marie Beebe and Robert M. Senkewicz

This groundbreaking collection presents an insider's view of Spanish and Mexican California. Taken from the writings of early explorers and residents, *Lands of Promise and Despair* offers a surprising and passionate image of California before the gold rush.

November Grass

By Judy Van der Veer, foreword by Ursula K. Le Guin

This novel transports readers to the coastal hills of San Diego County, where hawks and jays, calves and kittens, and an assortment of back-country eccentrics bring clarity to questions of birth, death, and love.

The Shirley Letters: From the California Mines, 1851–1852

By Louise Amelia Knapp Smith Clappe, edited with an introduction by Marlene Smith-Baranzini

With the grandeur of the Sierra Nevada as background, this classic account presents a picture of the gold rush that is at times humorous, at times empathetic, and always engaging.

Unfinished Message: Selected Works of Toshio Mori

Introduction by Lawson Fusao Inada

This collection features short stories, a never-before-published novella, and letters from a pioneer Japanese American author.

Unfolding Beauty: Celebrating California's Landscapes

Edited with an introduction by Terry Beers

The astounding beauty of California is reflected in this collection of pieces by John Muir, John Steinbeck, Wallace Stegner, Jack Kerouac, Joan Didion, and sixty-four other outstanding writers.

If you would like to be added to the California Legacy mailing list, please send your name, address, phone number, and email address to:

California Legacy Project
English Department
Santa Clara University
Santa Clara, CA 95053

For more on California Legacy titles, events, or other information, please visit www.californialegacy.org.

A CALIFORNIA LEGACY BOOK

Santa Clara University and Heyday Books are pleased to publish the California Legacy series, vibrant and relevant writings drawn from California's past and present.

Santa Clara University—founded in 1851 on the site of the eighth of California's original 21 missions—is the oldest institution of higher learning in the state. A Jesuit institution, it is particularly aware of its contribution to California's cultural heritage and its responsibility to preserve and celebrate that heritage.

Heyday Books, founded in 1974, specializes in critically acclaimed books on California literature, history, natural history, and ethnic studies.

Books in the California Legacy series appear as anthologies, single author collections, reprints of important books, and original works. Taken together, these volumes bring readers a new perspective on California's cultural life, a perspective that honors diversity and finds great pleasure in the eloquence of human expression.

Series editor: Terry Beers
Publisher: Malcolm Margolin
Advisory committee: Stephen Becker, William Deverell, Peter Facione, Charles Faulhaber, David Fine, Steven Gilbar, Dana Gioia, Ron Hansen, Gerald Haslam, Robert Hass, Jack Hicks, Timothy Hodson, James Houston, Jeanne Wakatsuki Houston, Maxine Hong Kingston, Frank LaPena, Ursula K. Le Guin, Jeff Lustig, Tillie Olsen, Ishmael Reed, Alan Rosenus, Robert Senkewicz, Gary Snyder, Kevin Starr, Richard Walker, Alice Waters, Jennifer Watts, Al Young.

Thanks to the English Department at Santa Clara University and to Regis McKenna for their support of the California Legacy series.